ESSENTIALS OF
MATHEMATICAL
THINKING

TEXTBOOKS in MATHEMATICS

Series Editors: Al Boggess and Ken Rosen

PUBLISHED TITLES CONTINUED

INVITATION TO LINEAR ALGEBRA
David C. Mello

LINEAR ALGEBRA, GEOMETRY AND TRANSFORMATION
Bruce Solomon

MATHEMATICAL MODELLING WITH CASE STUDIES: USING MAPLE™ AND MATLAB®, THIRD EDITION
B. Barnes and G. R. Fulford

MATHEMATICS IN GAMES, SPORTS, AND GAMBLING—THE GAMES PEOPLE PLAY, SECOND EDITION
Ronald J. Gould

THE MATHEMATICS OF GAMES: AN INTRODUCTION TO PROBABILITY
David G. Taylor

A MATLAB® COMPANION TO COMPLEX VARIABLES
A. David Wunsch

MEASURE AND INTEGRAL: AN INTRODUCTION TO REAL ANALYSIS, SECOND EDITION
Richard L. Wheeden

MEASURE THEORY AND FINE PROPERTIES OF FUNCTIONS, REVISED EDITION
Lawrence C. Evans and Ronald F. Gariepy

NUMERICAL ANALYSIS FOR ENGINEERS: METHODS AND APPLICATIONS, SECOND EDITION
Bilal Ayyub and Richard H. McCuen

ORDINARY DIFFERENTIAL EQUATIONS: AN INTRODUCTION TO THE FUNDAMENTALS
Kenneth B. Howell

PRINCIPLES OF FOURIER ANALYSIS, SECOND EDITION
Kenneth B. Howell

REAL ANALYSIS AND FOUNDATIONS, FOURTH EDITION
Steven G. Krantz

RISK ANALYSIS IN ENGINEERING AND ECONOMICS, SECOND EDITION
Bilal M. Ayyub

SPORTS MATH: AN INTRODUCTORY COURSE IN THE MATHEMATICS OF SPORTS SCIENCE AND SPORTS ANALYTICS
Roland B. Minton

TRANSFORMATIONAL PLANE GEOMETRY
Ronald N. Umble and Zhigang Han

TEXTBOOKS in MATHEMATICS

ESSENTIALS OF MATHEMATICAL THINKING

Steven G. Krantz

CRC Press
Taylor & Francis Group
Boca Raton   London   New York

CRC Press is an imprint of the
Taylor & Francis Group   an **informa** business
A CHAPMAN & HALL BOOK

CRC Press
Taylor & Francis Group
6000 Broken Sound Parkway NW, Suite 300
Boca Raton, FL 33487-2742

© 2018 by Taylor & Francis Group, LLC
CRC Press is an imprint of Taylor & Francis Group, an Informa business

No claim to original U.S. Government works

International Standard Book Number-13: 978-1-1380-4257-5 (Hardback)
International Standard Book Number-13: 978-1-1381-9770-1 (Paperback)

Visit the Taylor & Francis Web site at
http://www.taylorandfrancis.com

and the CRC Press Web site at
http://www.crcpress.com

Printed and bound in Great Britain by
TJ International Ltd, Padstow, Cornwall

To the memory of Ed Nelson.

Table of Contents

Preface

It is becoming increasingly clear that mathematical thinking is essential to understanding the world around us. Police tell us that most crimes these days are committed with electronic devices (usually computers). Many if not most medical procedures are now robotically assisted. The carburation system of a modern automobile is all electronic. Geometric visualization is one of the most important aspects of modern technology.

And it is plain that the thinking behind all of the technology described in the last paragraph is mathematical. The `Google` search engine is based on a mathematical algorithm. Not just logic, but analysis and combinatorics and geometry and many other ideas from the mathematical sciences are critical parts of our world.

A college education consists of learning different modes of discourse. Every modern student should thus have some exposure to mathematical discourse. Whether or not the student has any intention of working in the mathematical sciences, he/she will need these modes of thought to survive in the modern marketplace.

And learning mathematical thought does not have be dry, desultory, and boring. It can be fun and exciting and challenging (in a productive fashion). It can relate to things that are meaningful to students—such as the paradigm behind a Google search, or the means by which a music CD is encoded, or the RSA encryption algorithm. In a course on mathematical thinking the student will obviously learn some mathematical skills, but he/she will also gain *critical thinking skills*. These are valuable life tools.

This book may be thought of as a book for nonmathematicians taking a mathematics course that is a breadth requirement for an undergraduate degree. Thus it has minimal mathematical prerequisites. It has very few equations. It has lots of pictures and lots of explanation. It is driven by examples. But those examples will lead students down new paths, and acquaint them with new paradigms of thought. They will stimulate and provoke and encourage. At the end of this journey, the student should feel that he/she has a new set of tools for attacking a variety of problems and situations. There should be a profound feeling of satisfaction.

A very important part of learning mathematical thinking is learning to write mathematics. The process of internalizing an idea and the process of figuring out how to write it down are closely linked. We exploit this observation by supplying on-the-fly exercises that force the student to *write out* an answer. Sometimes, when it is appropriate and not burdensome, we ask the student to provide a demonstration or a reason for a mathematical idea. We also ask the student to draw pictures.

The exercises in this book are of several types: **(a)** there are drill exercises, just as a means of getting students started in a topic, **(b)** there are more challenging problems that will require the student to really digest the key ideas, and **(c)** each section ends with an open-ended problem that will encourage students to talk to each other and collaborate.

The material in the first nine chapters of the book should be accessible to a broad audience with minimal background in mathematics. The last three chapters are more sophisticated. All readers will want to dip into these chapters, but only readers with some mathematical background will be able to work through the examples in any detail. In this way, we have been able to make the book more open-ended and appealing to a diverse group of readers with many needs and interests.

It is a pleasure to thank Lynn Apfel and Tim Davis for helpful remarks and suggestions for many different parts of this book. I thank the many fine reviewers that Taylor & Francis engaged to criticize my book. As always, my editor Robert Ross was supportive and proactive in the production of this book.

Steven G. Krantz
St. Louis, Missouri

Chapter 1

First Thoughts

1.1 What Is Mathematical Thinking?

It is natural to wonder how mathematical thought differs from other modes of cogitation. Does mathematical thought require special skills and special training? Does one need to work at it in the same way that one works to learn to play the piano? Does one need to practice? Does one require a teacher?

This book will show you that you can learn mathematical thought—by yourself—as you enjoy reading along. It will pose questions and set up puzzles and help you learn how to answer them. It will teach you to draw pictures and then to analyze those pictures in order to elicit the solution of a problem. It will teach you to think by analogy. It will teach you how to generate your own ideas.

The chief feature of mathematical thinking is that it is *logical*. Certainly there is room for intuition in mathematics, and even room for guessing. But, in the end, we understand a mathematical situation and/or solve a problem by being very logical. Logic makes the process dependable and reproducible. It shows that what we are producing is a verifiable truth.

1.2   How Does Mathematics Differ from Other Disciplines?

Mathematics is not like biology or physics or chemistry. A mathematician does not work in a laboratory. A mathematician does not attempt to describe and understand the world around us. A mathematician does not (at least not directly) respond to nature.

A mathematician deals with ideas. He/she may use a computer to help analyze those ideas. The mathematician may even build a physical model to aid in thinking. But, in the end, a solution to a mathematical problem comes from pure thought.

And, following the model set by Euclid of Alexandria 2500 years ago, our thoughts follow rather strict rules. We set up a collection of definitions (of terminology and ideas), then we record certain axioms (these are the "rules of the game"), and finally we derive new truths from those axioms using rules of logic.

Along the way we may do a considerable amount of calculating. We may work many examples. We may "try" a number of different things. But, when the situation is sorted out and understood and the solution to the problem is well in hand, then we record it once and for all following the logical model described in the last paragraph.

1.3   A Sample Problem

Our sample problem will be analyzed using mathematical-style thinking. But it will not involve equations and it will not involve calculations. It will just involve *logic*. But that is the essence of mathematics.

Problem 1.1 *You are on an island that is populated by two types of people: truth tellers and liars. See Figure 1.1. When asked a Yes–No question, a truth teller always tells the truth and a liar always lies. There is no visual method for telling a truth teller from a liar. What single question could you ask anyone that you meet on the island to determine whether that person is a truth teller or a liar?*

Figure 1.1: A Liar and a Truth Teller.

Solution: If you ask a direct question such as "Are you a truth teller?" then a truth teller will answer "Yes" and a liar (who must lie) will also answer "Yes." You will get a similar result if you ask "Are you a liar?" Thus an elementary, direct question provides no basis for differentiation.

Therefore a compound question, such as a conditional, or an "or" question, or an "and" question is called for. One of the things that we learn in a basic logic course is that any question that is of one of these three types can be reformulated as a question of any one of the other three types (see [KRA1]). We concentrate on formulating an "if-then" question that will do the job.

The question that we formulate could be of the form "If it is raining then what would you say to ..." or "If you are a Doctor of Letters then what would be your answer to ...." However it is clear that these conditions have nothing to do with the matter at hand.

Probably more relevant would be a question of the form "If you were a truth teller then what would you say to ...." Likewise the concluding part of the question ought to have something to do with the problem that we are trying to solve. We now try the question

> **If you are a truth teller then how would you answer the question, "Are you a liar?"**

Now we analyze how the two different types of island inhabitants would answer this question.

Obviously a truth teller would answer the question "Are you a liar?" by saying "No." If you pose the displayed question to a truth teller, then he will report truthfully on the answer just given, so he/she will say "No."

A liar can think just as clearly as a truth teller. He/she also knows that a truth teller, if asked whether he/she is a liar, will say "No." But the liar must lie. So he/she will say "Yes."

Thus we have found a question to which a truth teller will always answer "No" and a liar will always answer "Yes." This certainly gives a means for differentiating truth tellers from liars, and answers our question.

Exercise: Consider an island with truth tellers, liars, and equivocal people. Can you cook up a question that will enable you to distinguish these three types of people?

Exercise: There are two identical doors. Behind one door is the passage to heaven. Behind the other door is the passage to hell. In front of each door is a guard. One of these guards is a truth teller and one of the guards is a liar. You cannot tell visually which is which. What question can you ask one of the guards to determine which door is which? [This question is due to Raymond Smullyan.]

Chapter 2

Diverse Mathematical Thoughts

2.1 A Fraction of the Time

The UCLA campus is located in Westwood, California. Westwood is one of the prime real estate locations in Southern California. Few faculty can afford to live in Westwood. Westwood Village, adjacent to the campus, is the home of a number of upscale boutiques and cafes and shops. And also to a branch of the Bank of America. Every Spring, Bank of America puts an ad in the UCLA student paper encouraging students to apply for a job at B of A. And this ad always includes a footnote that says, "Must be able to add fractions."

Ahem. Well, it is a sad but true fact—which if you have ever taught mathematics you know all too well—that students have trouble adding fractions. Americans in general have trouble with fractions. Professionals who spend their time thinking about K–12 education will tell you that teaching fractions is a *big* problem. The true nature of this problem is reflected in the following tale of the burger wars.

There has been a "burger war" going on in the United States for several decades now. McDonald's is battling it out with Burger King, Wendy's, A & W Rootbeer, and several other chains. They all want your burger dollar. And they are willing to resort to a bizarre variety of ploys to get it.

One of the most vivid arithmetic failings displayed by Americans occurred in the early 1980s, when the A & W restaurant chain released a new hamburger to rival the McDonald's Quarter Pounder. With a third-pound of beef, the A& W burger had more meat than the Quarter Pounder; in taste tests, customers preferred A & W's burger. And it was the same price. A lavish A & W television and radio marketing campaign cited these benefits. Yet, instead of leaping at the great value, customers snubbed it. Only when the company held customer focus groups did it become clear why. The Third Pounder presented the American public with a test in fractions. And we failed. Misunderstanding the value of one-third, customers believed they were being overcharged. Why, they asked the researchers, should they pay the same amount for a third of a pound of meat as they did for a quarter-pound of meat at McDonald's? The "4" in "1/4," larger than the "3" in "1/3," led them astray. But is this depressing tale true? For verification of this bizarre story, refer to the 2007 memoir of real estate tycoon A. Alfred Taubman, who purchased the A & W chain back in the early 1980s. The author of a *New York Times* piece on the matter—which is adapted from her new book, *Building a Better Teacher*—cited it as her source on `Twitter`.

Taubman reports as follows. "Of course, not all my creative efforts to redefine and reenergize A & W were successful. In fact, one experience in particular still leaves a very bad taste in my mouth. We were aggressively marketing a one-third-pound hamburger for the same price as a McDonald's Quarter Pounder. But despite our best efforts, including first-rate TV and radio promotional spots, they just weren't selling. Perplexed, we called in the renowned market research firm Yankelovich, Skelly, and White to conduct focus groups and competitive taste tests. Well, it turned out that customers preferred the taste of our fresh beef over traditional fast-food hockey pucks. Hands down, we had a better product. But there was a serious problem. More than half of the participants in the Yankelovich focus groups questioned the price of our burger." "Why," they asked, "should we pay the same amount for a third of a pound of meat as we do for a quarter-pound of meat at McDonald's? You're overcharging us." Honestly. People thought a third of a pound was less than a quarter of a pound. After all, 3 is less than 4!

Several years ago, when gasoline was more than $4 per gallon, one station in the Midwest started selling gasoline by the liter for $1.25. People just flocked to the station, and were happy as clams to be getting gas for just $1.25. When a TV station interviewed the customers and asked them whether they realized that they were only getting a liter, or just over 1/4 of a gallon, for $1.25, the customers were perplexed. A typical reply was, "I don't care what units they are using to sell us the gas. It's only $1.25!!!"

As we have noted elsewhere, much of life is mathematical. It is worthwhile to understand a little mathematics so that we can convince ourselves that we know what we are doing.

Exercise: Part of this phenomenon is psychological. People are familiar with the concept of 1/2 pound. They *know* that 1/2 pound is less than 1 = 1/1 pounds. But they are much less familiar with 1/3 pound and 1/4 pound. Imagine that you are a schoolteacher. How would you explain these concepts to your students? How would you make them understand that 1/4 is less than 1/3?

2.2 How to Swindle on the Stock Market

For the most part, if you want to succeed on the stock market, then you have to know a lot about different commodities and you have to understand how things are traded. But there are certain tricks.

Imagine that you wake up one morning and you receive a letter (or perhaps an email) from some stock guy in New York City. It says

> I know a lot about the stock market. I know when it will go up and when it will go down. For instance, next Wednesday, International Xolotl will go up.

You don't think too much about it. But, sure enough, you check the stock market next Wednesday and International Xolotl has gone up. Kowabunga.

A few days later you receive another letter or email from this same guy that claims that, the following Tuesday, National Potrzebie will go down. Getting ever more curious, you check the stock market next Tuesday and, sure enough, National Potrzebie has gone down. Something must be going on here.

You get four more communications from this New York fellow, one per week. And each time he makes a prediction about the stock market. And each time he is right.

How could this be? This guy must have the stock market sewn up. He must have the whole situation wrapped around his little finger. But then the plot thickens.

In the seventh week he sends you a letter or email saying

> OK. Now you know what I can do. So now I am going to sell you my next prediction for $5,000.

Well, $5,000 is a lot of money. But this guy is infallible. How can anything go wrong? You pay him $5,000 and invest $20,000 on whatever he recommends (play the bull market if he predicts the stock market will go up and play the bear market if he predicts the stock will go down) and you are bound to come out a winner. There is practically no risk. It certainly beats playing the roulette wheel in Vegas (but see Section 7.3). So you send him the $5,000 and play the stock that he recommends and what happens?

Who knows? He certainly doesn't care, and you will never hear from him again. But he got $5,000 from you and from several other suckers as well. So he made out like a bandit. What is going on here?

What is interesting is that you played the game straight from beginning to end. You made no logical input into the game. You only observed and reacted in a logical way. And the guy sending you the prescient messages was also playing it straight. The scam is that you did not realize what you were reacting to.

The way that the New Yorker played the game is this. In the first week he wrote to 640 people. He told half of them that International Xolotl would go up. And he told half of them that International Xolotl would go down. Then he watched the board to determine what International Xolotl actually did. Let us say that it went up. So he discards the 320 names to whom he predicted that it would go down. And concentrates on the remaining 320 to whom he would predicted it would go up. Because they are now the believers.

Now, in the next week, he writes to 160 of the remaining suckers and tells them that National Potrzebie will go up. And he writes to the other 160 and tells them that it will go down. With half of those he will be right. He keeps those 160 and discards the others.

So now it is the third week and he is working with 160 marks. He tells half of them that some stock will go up and half that it will go down. So, going into the fourth week, he has 80 saps on the towline.

In the fourth week he plays the game again and whittles his crew down to 40.

In the fifth week he pares the group down to 20.

And, after six weeks, he has 10 true believers who, checkbooks in hand, are each ready to fork over $5,000 for the big kill.

At that point Mr. New York does not really care what prediction he sends to these 10 targets. He has his $50,000 and he is out of here. Has he done anything illegal? Well, certainly he has committed some kind of fraud, but it would be quite difficult to prosecute. For how is what he did much different from what your stockbroker does on a daily basis?

Exercise: Imitate the scheme described in this section to come up with a scheme for predicting what grades a particular student will get on his/her next report card.

2.3   The Bible Code

In 1998, Michael Drosnin published a book called *The Bible Code* [DRO1]. In this book, he claims that there are certain messages hidden in the text of the Old Testament of the Bible. One can reveal these messages by looking at equally spaced letters in the text (such an item is commonly referred to as an ELS, or *equidistant letter sequence*). Using this technique, people have found the word "Torah" encoded in the Bible. They have also found the longer message

> Destruction I will call you; cursed is Bin Laden and revenge is to the Messiah.

One should keep in mind that the Old Testament of the Bible was originally written in Hebrew. Hebrew of course uses a different alphabet from the Roman alphabet that we use. It is also written from right to left instead of left to right. So any version of the Bible that we may read will be a translation, and there are many of those. Hardcore proponents of the Bible code work only from the original Hebrew, which was handed down by Moses and therefore is considered to be definitive.

I was reading a book the other day on school safety. It contained the passage

Once born, alarm management can go awry.

Let us use Drosnin's idea, and look at ELS examples that are spaced four apart. In the following display, the selected letters are typeset in boldface:

Once **b**orn, **alarm** management can go awry.

My goodness. We have uncovered the hidden message "Obama." What is the author, or perhaps God, trying to tell us here? That President Barack Obama favors school safety? That he does *not* favor school safety? Or something else?

Drosnin's book asserts that the codes uncovered in the Bible can be used to predict future events. In particular, some Bible code avatars have uncovered the encoded names and birthdates of rabbis who were born many hundreds of years after the Old Testament was written.

The exact order of consonant letters represented in the original Hebrew Masoretic Text was only finalized in its current form in the first century. It is known from earlier versions of the Old Testament, such as the Dead Sea Scrolls, that the number of letters was not constant before this. The Bible Code theory does not account for these variations.

Interestingly, one of the main scientific objections to the Bible Code theory concerns noise. If the data chosen for ELS experiments are "cooked" before the experiment is conducted, then similar patterns may be found in all different kinds of texts (not just the Old Testament). In a long text, there are so many starting points and so many different skip patterns that it is no great surprise that many interesting words will appear. It is possible to "tune" an ELS experiment to obtain results which appear to exhibit patterns that overcome any noise that is present.

The Drosnin book and related experiments performed by statisticians and others have generated considerable controversy. One group of researchers extracted the first 78,064 letters from the Tolstoy novel *War and Peace* (this is the same number of letters as are in the Book of Genesis in the Old Testament of the Bible). And they ran a computer search on this sample for ELS. And they found about the same number of hits. This suggests that there is nothing special about the Book of Genesis.

Others have accused Drosnin of mixing different versions of Hebrew (some of which have no vowels), and also using various spellings, to enhance his results. Also it has been said that Drosnin skews the interpretation of his findings to support the predictive claims that he makes.

Some of Drosnin's predictions are just plain wrong. In the second volume of his book [DRO2], he predicts an atomic holocaust that will end the world. He also predicts that Libya will develop atomic weapons and give them to terrorists who will attack the United States. Finally Drosnin predicts that Yasser Arafat will be assassinated by Hamas gunmen. In point of fact, Arafat died in 2004 of natural causes.

The discussion of the merits of the Bible Code theory continue. It illustrates how easily mathematics can be used to mislead people.

Exercise: Given any sentence with 50 characters in it (50 letters from the alphabet), there is some scheme that will find that sentence encoded in the Old Testament. It need not be an equidistant letter sequence. It could be some other polynomially coded encryption method. Explain what this might mean and give an example.

2.4 Winning on a Game Show

The next problem has received a considerable amount of publicity in the last few years. It was inspired by the television game show *Let's Make a Deal*. The nature of the game show (a bit over-simplified) is as follows. The contestant is faced with three doors. See Figure 2.1. He/she knows that behind one door is a very desirable prize—say a fancy car. Behind the other two doors are rather pesky and undesirable items—say that a goat is behind each.[1] The contestant is to pick a door (blind), and is awarded the prize that is behind the door. But the game show host, Monty Hall, teases

---

[1]We note that, in Afghanistan, perhaps the goats would be considered to be the desirable prizes. But we are assuming that the game is played in New York City.

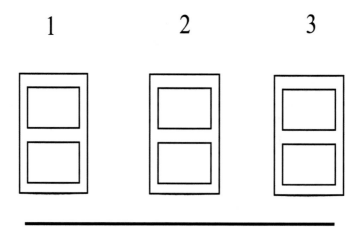

Figure 2.1: Three doors on *Let's Make a Deal*.

and cajoles and bribes the contestant, encouraging the contestant to change his/her mind and forcing the contestant to become confused over which is the most desirable door.

What has become known as the "Monty Hall" problem is this: The contestant picks a door. For the sake of argument, we say that he/she has picked Door 3. Before the door is opened, revealing what is behind it, Monty Hall says "I will now reveal to you what is behind one of the other doors." Door 1 is opened and there stands a goat. See Figure 2.2. Then Monty Hall says "Would you like to change *your* door selection?" Very interesting.

Clearly the contestant will not pick the door that Monty Hall has already opened (Door 1), since that has a goat behind it. So the issue is whether the contestant will switch from the currently selected door (Door 3) to the remaining door (Door 2, the one that the contestant has not chosen and Monty Hall did not open). A naive approach would be to say there is an equal probability for there to be a goat behind the remaining door (Door 2) and behind the door that the contestant has already selected (Door 3)—after all, one door has a goat and one has a car. What is the point of switching? However this naive approach does not take into account the fact that there are two distinct goats. A more careful analysis of cases occurs in our solution to the problem, and reveals a surprising answer.

Let us use a case-by-case analysis to solve the Monty Hall problem. We denote the goats by G_1 and G_2 (for goat one and goat two) and the car by

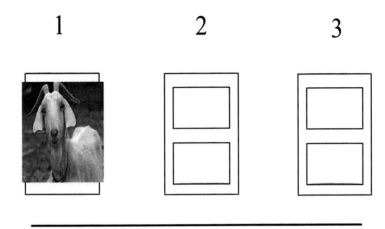

Figure 2.2: Door 1 has a goat.

C. For simplicity, we assume that the contestant will always select Door 3. We may not, however, assume that Monty Hall always reveals a goat behind Door 1; for there may not be a goat behind Door 1 (it could be behind Door 2). Thus there are several cases to consider:

Door 1	Door 2	Door 3
G_1	G_2	C
G_2	G_1	C
G_1	C	G_2
G_2	C	G_1
C	G_1	G_2
C	G_2	G_1

As we know from Section 10.2 below, there are $6 = 3!$ possible permutations of three objects. That is why there are six rows in the array.

1. In the first case, Monty Hall will reveal a goat behind either Door 1 or Door 2. It is *not* to the contestant's advantage to switch (because he/she has already decided on Door 3, which has the car), so we record **N**.

2. The second case is similar to the first, it is not to the contestant's advantage to switch, and we record **N**.

3. In the third case, Monty Hall will reveal a goat behind Door 1, and it *is* to the contestant's advantage to switch. We record **Y**.

4. The fourth case is like the third, and it *is* to the contestant's advantage to switch. We record **Y**.

5. In the fifth case, Monty Hall will reveal a goat behind Door 2. It *is* to the contestant's advantage to switch, so we record **Y**.

6. The sixth case is like the fifth, it *is* to the contestant's advantage to switch, and we record **Y**.

Observe that the tally of our case-by-case analysis is four **Y**'s and just two **N**'s. Thus the odds are two against one in favor of switching after Monty Hall reveals the goat.

Exercise: Answer the Monty Hall question if there are 4 doors instead of 3 doors and 3 goats instead of 2. What happens in this last case if you assume there are 2 goats instead of 3?

2.5   Cutting the Cake

Imagine presenting two kids with a cake, and they are to divide it up (Figure 2.3). Now both of these kids are *very* hungry, and they each want to maximize the amount of cake that they will enjoy. Also each has certain preferences. One kid likes cake with a chocolate frosting, and the other kid likes cake with a vanilla frosting (the cake in question has both). How to divide the cake?

A classical answer to this question, which in fact goes back millennia in our history, is that you let one kid cut the cake, and let the other kid choose his/her piece. This is a good answer, and it is not a frivolous one. In fact international diplomats at the negotiation table use this strategy frequently.

In case there are more than two people among whom to divide the cake, or more than two criteria for identifying desirable pieces (other criteria besides

Figure 2.3: A cake for cutting.

rosting), or the two people value the two criteria with different proportion, hen the plot thickens.

These ideas are best illustrated with an example. Let us suppose that George and Mary are going to divide the cake. Each assigns a numerical value o chocolate and a numerical value to vanilla. The sum of the values for each person must be 10. So we shall take it that Mary values the chocolate at 9 and the vanilla at 1 while George values the chocolate at 6 and the vanilla at 4. We want to divide the cake so that each person achieves a value of at east 5 (which is half of the total value of 10).

As a first try, we might give all of the vanilla and 4/9 of the chocolate o George. And give 5/9 of the chocolate to Mary. One then calculates the value for George to be

$$1 \cdot 4 + \frac{4}{9} \cdot 6 = \frac{20}{3} \approx 6.66 \, .$$

And the value for Mary is

$$0 \cdot 1 + \frac{5}{9} \cdot 9 = 5 \, .$$

This may be considered as a satisfactory outcome, since both kinds have achieved a value of at least 1/2.

But in fact one can do better. Suppose we give half the chocolate and half the vanilla to George. And we give half the chocolate and half the vanilla to

Mary. Then the value for George is

$$\frac{1}{2} \cdot 6 + \frac{1}{2} \cdot 4 = 5 \,.$$

And the value for Mary is

$$\frac{1}{2} \cdot 9 + \frac{1}{2} \cdot 1 = 5 \,.$$

This seems like an even more fair division of the cake. It is common to call this analysis a *proportional division*.

A particularly interesting form of the problem is *envy-free* cake division. This is the program to divide a cake (among three or more people) so that nobody thinks that anybody else's share is better than his/her own. This problem has been solved only recently for three participants. It is still open for four or more participants.

We shall describe here a solution for three cake eaters. It is actually a tricky, iterative procedure devised by Conway and Selfridge (unpublished). Let us call the three participants Alice, Bob, and Charlie. We will begin by creating an envy-free (EF) division in which there are some extra "trimmings." Later on we shall divide up the trimmings in an equitable fashion.

Stage One: Divide Cake with Trimmings   We begin by having Alice divide the cake into "equal thirds" *by her measure*. Now we have Bob examine the three pieces that Alice has created and trim the largest piece (in his perception) so that it is now (in his view) equal in value to the second-largest piece. We shall denote this trimmed piece by T. Put the trimmings aside, and we shall deal with them later.

Stage Two: Cyclic Choice   Now we have the players go in order: Charlie, Bob, Alice. They each pick the slice that they like best, with the proviso that Bob must pick slice T if Charlie has not already selected it. We claim that this setup is EF.

It is EF for Charlie because Charlie chooses first. It is EF for Bob since he has two equally favorite pieces. It is EF for Alice since she split the cake equally in the first place. The only slice that she might not want is T (since that is a trimmed-down version of one of her slices), but the setup guarantees that she will not have to take T.

Stage Three: Dealing with the Trimmings   It is now important to note that Alice will remain EF no matter how much cake we give to the person who received T—just because she is the one who originally cut the cake into equal thirds. But she could conceivably envy the person who did not get T, so we cannot ignore Alice.

Now whichever of Bob or Charlie did not get piece T should now cut the trimmings from Stage One into thirds. The other of these two (the one who *did* get piece T) chooses first, then Alice, and the person who is now cutting goes last. We claim that this stage of the game is also EF.

It is obviously EF for the person who chooses first, since he/she has free choice. It is EF for Alice, since she certainly will not envy the person who got T and she takes her turn before the other player. It is EF for the person who goes last since he/she did the cutting and made the three equal pieces.

It is worth noting here that the additivity of the procedure is crucial. If you do not envy someone's Stage One slice and you do not envy their Stage Two and Stage Three slices, then you do not envy their combination.

Exercise: Discuss the situation where two people are sharing a cake, but one of them is dishonest and steals 20% of the other one's cake.

2.6   A Lesson in Map Coloring

In the mid-nineteenth century some British college students came up with the following enticing question. Suppose that you have a map consisting of various countries. No bodies of water—just countries. And the definition of a country is that it is a connected land mass with no holes in it. We would not allow Italy because it has the Vatican in the middle of it. We would not allow the United States because it has Alaska and Hawaii. But we would allow Germany and France. The question is how many colors it takes to color this map. The only rule about coloring is that, if two countries share an edge, then they must be of a different color. See Figure 2.4, in which B stands for "blue," R stands for "red," Y stands for "yellow," and G stands

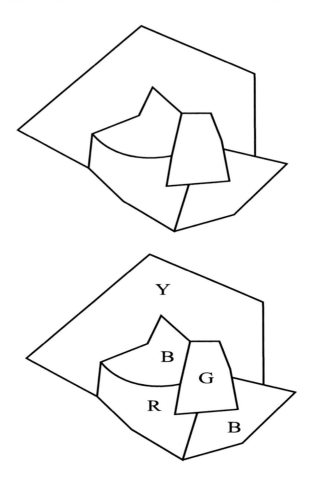

Figure 2.4: A map and a possible coloring.

for "green."

Modern mathematics is a rich and complex tapestry of ideas that have evolved over thousands of years. Unlike computer science or biology, where the concept of truth is in a constant state of flux, mathematical truth is permanent. Ideas that were discovered 2000 years ago by Pythagoras are still valid today. Proofs (also millennia old) in the style of Euclid are as valid today as when they were first created.

As a result, modern mathematics can be quite complex and technical. It requires someone with considerable advanced training to understand the current problems, much less solve them. As an instance, it was a huge event in the world of mathematics when, around 2002, it was announced that

Grigori Perelman had proved the Poincaré conjecture. This is a very famous problem posed in 1904 by Henri Poincaré. It is a question about the shape of the universe. It has profound implications both for pure mathematics and for cosmology.

Most mathematicians, when they think they have solved a big problem, write up the result carefully and in great detail and circulate it (either on a Web site called a preprint server or by email) to colleagues. But not Perelman. He put four brief papers on the preprint server arXiv which sketched his ideas. But he left out most of the details, and also had several shaky leaps and passages which read "I'm not quite sure how to do this but it probably works for the following reason." Let me tell you that this is not the way that mathematics is usually done. But Perelman had considerable credibility because he had done some excellent work in the past.

Many mathematicians, including Bruce Kleiner, John Lott, John Morgan, Gang Tian, Zhu Xiping, and Huai-Dong Cao jumped in and labored (for years!) to fill in the details of Perelman's arguments. Morgan and Tian actually published a 450-page book with all of the detailed arguments. It is now widely accepted that Perelman has proved the Poincaré conjecture. He was awarded the Fields Medal (the highest award in mathematics) and also the first Clay Millennium Prize of $1 million. He declined both of these prizes.

I tell you this story about the Poincaré conjecture because it illustrates the recondite nature of mathematics. I frankly cannot tell you the precise statement of the Poincaré conjecture because it is too technical. But the events that I have described here shook the mathematical world to its core. This was all that anyone was talking about for several years. Perelman himself was so upset by the way that events unfolded that he has quit his professorship and has terminated his mathematical career. He now spends his time taking long walks and listening to opera. And taking care of his mother. Rumor has it that he has recently moved to Sweden and now works for a high-tech startup company.

I'm trying to set the stage here for telling you how charming it is when we can find problems that *anyone* can understand, but that still resist the best efforts of the world's great experts. In fact the problem that I am going to discuss with you in this section could be explained to your grandmother or your little sister. And, likely as not, they would pick up a pencil and try

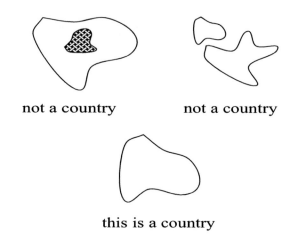

not a country   not a country

this is a country

Figure 2.5: What is a country?

to solve it.

The problem that we wish to discuss today is charming and simple. It is appealing because it is geometric, and it has an interesting and unusual genesis. In 1852 Francis W. Guthrie, a graduate of University College London, posed the following question to his brother Frederick:

Imagine a geographic map on the Earth (i.e., a sphere) consisting of countries only—no oceans, lakes, rivers, or other bodies of water. The only rule is that a country must be a single contiguous mass—in one piece, and with no holes—see Figure 2.5.

As cartographers, we wish to *color* the map so that no two adjacent countries (countries that share an edge) will be of the same color (Figure 2.6). How many colors should the map-maker keep in stock so that he can be sure he can color any map that may arise? This is the famous *four-color problem.*

2.6.1   Analysis

It is not difficult to write down an example of a map that surely needs 4 colors. Examine Figure 2.7. Each of the countries in this figure is adjacent to each of the others. There are four countries, and they all must be of a different color.

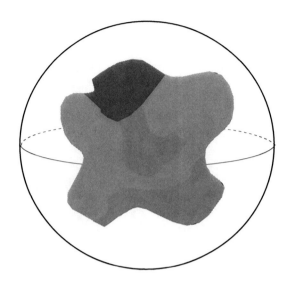

Figure 2.6: A typical map and its coloring.

Is there a map that will require 5 colors? Mathematicians at the best universities beat their collective heads against this question for many decades.

The eminent geometer Felix Klein (1849–1925) in Göttingen heard of the problem and declared that the only reason the problem had never been solved is that no capable mathematician had ever worked on it. *He*, Felix Klein, would offer a semester-long class, the culmination of which would be a solution of the problem. *He failed.* And Klein is not the only famous mathematician who fell on his face over this problem. It baffled the best minds in mathematics for more than 100 years.

In 1879, A. Kempe (1845–1922) published a solution of the four-color problem. That is to say, he showed that any map on the sphere could be colored with four colors. Kempe's proof stood for eleven years. Then a mistake was discovered by P. Heawood (1861–1955). Heawood studied the problem further and came to a number of fascinating conclusions:

- Kempe's proof, particularly his device of "Kempe chains" (a sequence of countries that alternates between just two colors), *does* suffice to show that any map can be colored with at most 5 (*not* 4) colors. We say that the *chromatic number* of the sphere is *at most* 5, but it *could*

Figure 2.7: A map that requires four colors.

be 4.

- Heawood found a formula that gives an estimate for the chromatic number of any surface that is geometrically more complicated than the sphere.

Here is how to understand Heawood's idea. It is known that any surface in space is geometrically equivalent to a sphere with handles attached. See Figure 2.8. The number of handles is called the *genus,* and we denote it by g. The Greek letter chi ($\chi(g)$) is the chromatic number of the surface—the least number of colors that it will take to color any map on the surface with genus g.

Heawood's formula is

$$\chi(g) \leq \frac{1}{2}\left(7 + \sqrt{48g + 1}\right)$$

so long as $g \geq 1$.

The torus (see Figure 2.9) or donut surface is topologically equivalent to a sphere with one handle. Thus the torus has genus $g = 1$. Then Heawood's formula gives the estimate 7 for the chromatic number:

$$\chi(1) \leq \frac{1}{2}\left(7 + \sqrt{48 \cdot 1 + 1}\right) = \frac{1}{2}(7 + 7) = 7\,.$$

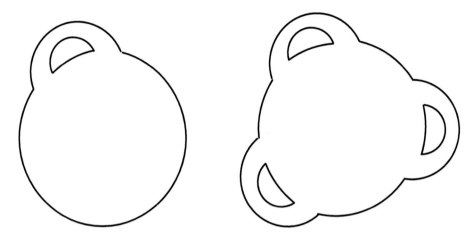

Figure 2.8: Any surface in space is a sphere with handles attached.

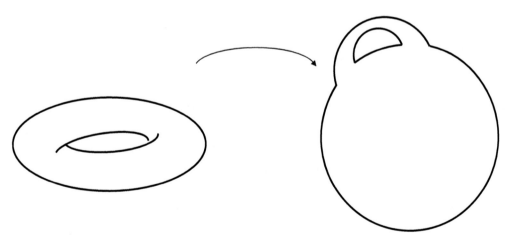

Figure 2.9: The torus.

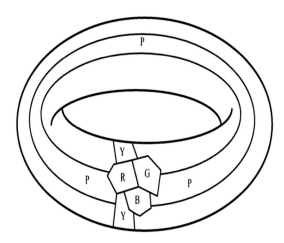

Figure 2.10: A map on the torus that requires five colors.

In fact one can experiment around a bit on the torus and determine fairly quickly that four colors will not suffice to color every map. Examine Figure 2.10 to see a map on the torus that requires five colors.

In fact seven definitely is the correct chromatic number for the torus— there really is a map that requires seven colors. Put in other words, there is a map with seven countries each of which touches the other. So they all must be of a different color. But it is rather difficult to draw such a map on a standard depiction of a torus such as the ones we have been examining. In order to tackle this problem we need a mathematician's trick. Geometric visualization is a big part of mathematics these days. There are many computer people who spend their time developing utilities for geometric visualization. We are here going to use a very elementary but effective trick of geometric visualization.

The idea here is that we are going to take a pair of scissors and cut the torus apart. With two cuts we can turn it into a rectangle with certain "identifications" of the edges. Having done that, we will be able to exhibit seven countries with all the adjacencies that we require.

We can show that this is the *right* (or best possible) estimate by first performing the trick of cutting the torus apart. See Figure 2.11. By cutting the torus around the small rotation and then across the large rotation, we render it as a rectangle with identifications.

With one cut, the torus becomes a cylinder; with the second cut it be-

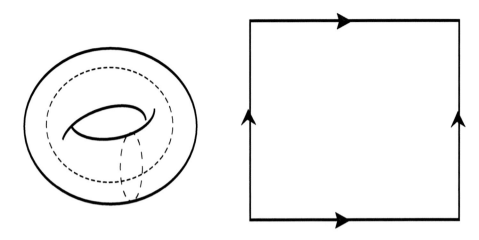

Figure 2.11: The torus cut apart.

comes a rectangle. The arrows on the edges indicate that the left and right
edges are to be identified (with the same orientation), and the upper and
lower edges are to be identified (with the same orientation). Now we can see
in Figure 2.12 how to color the torus (rendered as a rectangle). For clarity,
we call our colors "1," "2," "3," "4," "5," "6," "7."

We may see that there are seven countries shown in our Figure 2.12, and
every country is adjacent to (i.e., touches) every other. Take a few moments
to pick a country and check that it is adjacent to all the others!! Thus they
all must have different colors. This is a map on the torus that *requires* 7
colors; it shows that Heawood's estimate is sharp for this surface.

For the sphere with two handles (genus 2—see Figure 2.13), Heawood's
estimate gives an estimate of 8. Is that the best number? Is there a map
on the double torus that actually *requires* 8 colors? And so forth: we can
ask the same question for every surface of every genus. Heawood could not
answer these questions. Nor could anyone else.

2.6.2   Modern Developments

The late nineteenth century saw more alleged solutions of the four-color problem, many of which stood for as long as eleven years. Eventually errors were
found, and the problem remained open into the twentieth century.

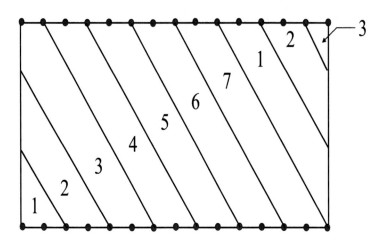

Figure 2.12: Seven colors on the torus.

Figure 2.13: The sphere with two handles.

What is particularly striking is that Gerhard Ringel (1919–2008) and J. W. T. Youngs (1910–1970) were able to prove in 1968 that all of Heawood's estimates, for the chromatic number of any surface of genus at least 1, are sharp. So the chromatic number of a torus is indeed 7. The chromatic number of a "double-torus" with two holes is 8. And so forth. I took a ten-week course from J. W. T. Youngs in which he explained their proof.

It was quite a long and elaborate argument. They had to have a method of constructing an *example* for each of infinitely many different surfaces. This reduced, after much recondite reasoning, to producing certain types of paths on certain graphs. This was the first graduate course that I ever took and it was very exciting for me. To be at a time and place where real live mathematics was taking place and being discovered was a formative experience for someone who was studying to be a mathematician.

One of the projects that I did for that class was to create a step-by-step outline of the entire proof. I did so with circles and arrows on a huge sheet of paper. It was really hard work to harness all that information and organize it. Youngs was thrilled when I presented it to him. I also had the pleasure of re-discovering some of Youngs's proofs on my own. That made me feel like a real mathematician, and I was pleased to get Youngs's approbation.

But the Ringel/Youngs proof, just like the Heawood formula, does not apply to the sphere. They could not improve on Heawood's result that 5 colors will always suffice for the sphere. So the 4-color problem remained unsolved.

Then in 1974 there was blockbuster news. Using 1200 hours of computer time on the University of Illinois supercomputer, Kenneth Appel (1932–2013) and Wolfgang Haken (1928–) showed that in fact 4 colors will always work to color any map on the sphere. Their technique is to identify 633 fundamental configurations of maps (to which all others can be reduced) and to prove that each of them is reducible to a simpler configuration. But the number of "fundamental configurations" was very large, and the number of reductions required was beyond the ability of any human to count. And the reasoning is extremely intricate and complicated. Enter the computer. In those days computing time was expensive and not readily available, and Appel and Haken certainly could not get a 1200-hour contiguous time slice for their work. So the calculations were done late at night, "off the record," during various down times. In fact, Appel and Haken did not know for certain whether the calculation would ever cease. Their point of view was this:

- If the computer finally stopped, then it will have checked all the cases and the 4-color problem was solved.

- If the computer never stopped, then they could draw no conclusion.

Well, the computer stopped. But the level of discussion and gossip and disagreement in the mathematical community did not. Was this really a proof? The computer had performed tens of millions of calculations. Nobody could ever check them all.

This is really a point worth pondering. Mathematics has a long tradition—going back to Euclid and even before him to Eudoxus and Thaetatus—of establishing mathematical facts using proofs. This was not always the case. Early mathematics was epistemological. Many questions of a mathematical sort arose from questions of land management, and the answers that were derived were of a very pragmatic nature. There were no axioms and no theorems and no proofs. When, for instance, you asked yourself whether a certain table would fit in your dining room, you did not prove a theorem; you just checked it out.

I cannot resist here telling a favorite story. William (Willy) Feller (1906–1970) was a prominent mathematician at Princeton University. He was one of the fathers of modern probability theory. Feller and his wife were once trying to move a large circular table from their living room into the dining room. They pushed and pulled and rotated and maneuvered, but try as they might they could not get the table through the door. It seemed to be inextricably stuck. Frustrated and tired, Feller sat down with a pencil and paper and devised a mathematical model of the situation. After several minutes he was able to *prove* that what they were trying to do was impossible. While Willy was engaged in these machinations, his wife had continued struggling with the table, and she managed to get it into the dining room.

When you wondered whether a certain amount of fence would surround your pasture, you did not seek a rigorous argument; you simply unrolled the fence and determined whether it did the job. In its earliest days, mathematics was intimately bound up with questions precisely like these. Thus mathematical thinking was almost inextricable from practical thinking. And that is how its adherents viewed mathematical facts. They were just practical

information, and their assimilation and verification was a strictly pragmatic
affair.

It was only over the course of centuries, and especially the efforts of Euclid
with his *Elements*, that the idea of an axiomatic structure, and theorems
derived from the axioms, came about. And this is the model we have been
following in mathematics now for 2500 years. Mathematicians believe in their
heart of hearts that what a mathematician does is as follows:

- Come up with an idea.

- Figure out why that idea is true.

- Devise a *proof* of that result.

- Write up the result, following the paradigm of Euclid, in a formal document (what mathematicians call a *paper*).

- Distribute that document to other mathematicians for them to *study*
  and to *validate*.

This is how mathematics has been practiced—around the world—for more
than two millennia. It is ingrained in all mathematicians that this is how
mathematics should be done. In particular, it is the body of mathematicians
the world over that vets and validates any new claim and any new proof.
This is how we did it with Perelman's proof of the Poincaré conjecture (and
without much help from Perelman). And this is how people wanted to do it
with the 4-color problem.

But Appel and Haaken did not play by the rules. One could read about
Appel and Haaken's algorithm, one could learn what their proof *strategy*
was. But one could not actually study the proof itself. This was humanly
impossible. And the situation really angered people. They felt that they had
been robbed. Appel and Haaken were invited to give a major invited address
at a big meeting of the American Mathematical Society to present their
results. The talk was heavily attended and people listened very carefully.
But at the end of the talk there was no applause. People were so puzzled
and offended by this situation that they did not know how to react to it.

But the University of Illinois mathematics department was no wilting
flower. It arranged for the postmark on all envelopes emanating from the

Illinois math department to read

Four colors suffice

It is easy to see, even without emotional involvement, why people would be perplexed by a computer proof. How do we know that there are no errors in the computer code? How do we know that there are no quantum-mechanical errors taking place in the central processing unit? How can we *check* a computer proof? How can we *understand* a computer proof? How can we *replicate* a computer proof? How can we *learn from* a computer proof? How can we *teach* a computer proof?

And that is really the bottom line. What a mathematician wants from a new mathematical result is *understanding*. He/she wants to learn something new. To apprehend some new techniques to add to the old toolkit. To feel that he/she has increased his/her substance as a mathematician. A computer proof does not achieve this effect.

It is worth noting that there has been a lot of water under the bridge since the Appel/Haaken proof. There is now a whole industry of checking computer proofs. There is specialized software that is dedicated to this task. There is also specialized software that can *discover* new mathematical facts. What you do is program a set of axioms into this software and it can fit them together (using rules of logic) to seek to discover new mathematical truths. And some striking new ideas have been discovered in this way.

I would be excessively naive to assert that now, after forty years, everyone has embraced computer proofs. But people have at least become accustomed to the fact that mathematicians use the computer to explore mathematics, to discover new facts, and sometimes to generate "proofs." I shall mention two instances that are dear to my heart.

David Hoffman at the University of Massachusetts and his team of collaborators (known as GANG, and short for Geometry, Analysis, and Computation in Mathematics and Applied Sciences) were interested in studying minimal surfaces. These are surfaces in space like soap bubbles which locally minimize area. There were questions of whether there existed certain

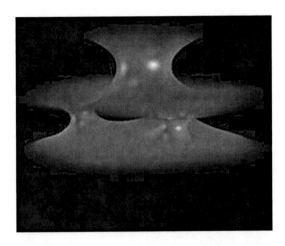

Figure 2.14: One of David Hoffman's minimal surfaces.

minimal surfaces with certain types of geometric complexity. David got the Department of Energy interested in the problem (because of possible applications to oil exploration), and they funded him heavily to buy a lot of computer equipment and to generate numerical solutions of the minimal surface differential equations together with graphics. This project was a great success, and they did discover some remarkable new minimal surfaces. Many of Hoffman's images are quite famous, and some of them are displayed in museums.

But here is the clincher. What Hoffman and his team did is to stare at the geometric images they created, discuss them, figure out what is going on, and then *write down a traditional mathematical proof.* This was terrific, because it gave their project solid credibility. The pretty graphics are some very nice icing on the cake, but nobody ever claimed that they were the essential mathematics. Nobody claimed that the pictures were a proof. They were just a tool used along the way. Look at Figure 2.14 to appreciate one of Hoffman's minimal surfaces.

The other project that I would like to mention is one that I engaged in with C. Eugene Wayne and Gordon Chen. We were studying a problem that grew out of the space station project. NASA wanted to create a 200-foot pole to connect the space shuttle (Figure 2.16) with the construction of the space station (Figure 2.17). And this poses very serious materials problems. If you just use standard materials to construct a 200-foot pole, then it will

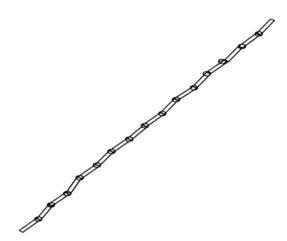

Figure 2.15: The pseudo-pole.

be too brittle. It cannot carry a load. So the idea was conceived to make the pole out of twenty 10-foot pieces, connected together with a special semi-stiff hinge and each hinge having a servo-motor. So we would have a twenty-piece "pseudo-pole" that could assume most any configuration. See Figure 2.15.

So what we had to do was study the fourth-order Euler–Bernoulli beam equations for such a system, and calculate the eigenvalue asymptotics. We were able to do this by hand for a two-piece pole. But three pieces was already too complicated, and twenty pieces seemed to be intractable. Enter the computer.

We used symbol manipulation software—specifically MACSYMA (a precursor of Mathematica)—to do the calculations for us in the case of a three-piece beam. And the computer was undaunted by the complexity of the problem. It just got in there and calculated.

We studied the computer output and were finally able to see what was going on. Pursuant to that study, we were able to do the calculations by hand—not just for the three-piece beam but also for the twenty-piece beam. So our project, in the end, was a success. And I can tell you somewhat proudly that I now have a published paper with a picture of the space shuttle in it (Figure 2.16).

Let us now return to the Appel-Haaken solution of the 4-color problem. It is around 1975 and the world has been endowed with this new computer-generated proof. What of it?

Figure 2.16: The space shuttle.

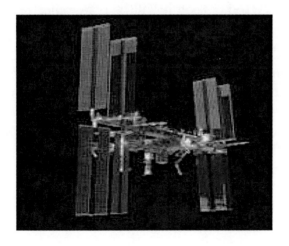

Figure 2.17: The space station.

But now the plot thickens. Because in 1975 a mistake was found in the proof. Specifically, there was something amiss with the algorithm that Appel and Haken fed into the computer. It was later repaired. The paper was published in 1976. The four-color problem was declared to be solved.

In a 1986 article, Appel and Haken point out that the reader of their seminal 1976 article must face

- 50 pages containing text and diagrams;

- 85 pages filled with almost 2500 additional diagrams; and

- 400 microfiche pages that contain further diagrams and thousands of individual verifications of claims made in the 24 statements in the main section of the text.

But it seems as though there is always trouble in paradise. Errors continued to be discovered in the Appel/Haken proof. Invariably the errors were fixed. But the stream of errors never seemed to cease. So, is the Appel/Haken work really a proof?

I cannot refrain from reiterating how traditionally oriented mathematicians view this matter. The old-school way of looking at things is that, once a theorem and proof have been written down and verified, then it is as solid as a rock. If it is true today then it will be true tomorrow. And generations hence will be able to learn from it just as we have learned from it today. Computer proofs are not like that.

Well, there is hardly anything more reassuring than another, independent proof. Paul Seymour and his group at Princeton University found another way to attack the problem [SEY]. In fact, they found a new algorithm that seems to be more stable. They also needed to rely on computer assistance. But by the time they did their work, computers were *much*, much faster. So they required much less computer time. In any event, this paper appeared in 1994. And in fact they solved a rather more general problem. People seem to be more comfortable with the Princeton approach to this problem.

2.6.3   Denouement

It is still the case that mathematicians are most familiar with, and most comfortable with, a traditional, self-contained proof that consists of a sequence of logical steps recorded on a piece of paper. We still hope that someday there will be such a proof of the four-color theorem. After all, it is only a traditional, Euclidean-style proof that offers the understanding, the insight, and the sense of completion that all scholars seek.

And there are new societal needs: theoretical computer science and engineering and even modern applied mathematics require certain pieces of information and certain techniques. The need for a workable device often far exceeds the need to be *certain* that the technique can stand up to the rigorous rules of logic. The result may be that we shall re-evaluate the foundations of our subject. The way that mathematics is practiced in the year 2100 may be quite different from the way that it is practiced today.

Exercise: The text discusses the sphere with two handles and what its chromatic number is. See what kinds of maps you can draw on this surface and how many colors you can force such a map to require.

2.7 The Complexity of Songs

One of the outstanding figures in modern computer science is Donald Knuth of Stanford University. Among his many accomplishments—including authoring the definitive work on modern theoretical computer science [KNU1] and creating the computer typesetting system TEX that is used today by mathematical scientists all over the world—is the authorship of a satirical piece called *The Complexity of Songs* [KNU2]. Let us explain.

A cornerstone of the modern ideas behind computer science is complexity theory. Given a task involving objects of size N, one would like to know how many machine steps it would take to perform that task. As an example, suppose you have a deck of N playing cards numbered 1 through N (Figure 2.18). You shuffle them so that they are in random order. How many steps will it take to put them back into their canonical order $1, 2, 3, \ldots, N$?

A simple answer to the question is this. You go through each card in sequence to find 1. That is your first card in the re-ordering. Then you go through the remaining $(N - 1)$ cards in sequence to find 2. That is your second card in the re-ordering. Then you go through the remaining $(N - 2)$

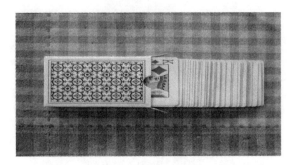

Figure 2.18: A deck of cards.

cards in sequence to find 3. Continue this procedure. After

$$N + (N - 1) + (N - 2) + \cdots + 3 + 2 + 1$$

steps you will have put the cards back into their natural order.

How many steps is this altogether? Call the displayed sum S. We can write

$$
\begin{aligned}
S &= N + (N-1) + (N-2) + \cdots + 3 + 2 + 1 \\
S &= 1 + 2 + 3 + \cdots + (N-2) + (N-1) + N.
\end{aligned}
$$

Now add each column. We notice that each column adds to $N + 1$, and there are precisely N columns to the right of the equal signs. So the result is

$$2S = \underbrace{(N + 1) + (N + 1) + \cdots + (N + 1)}_{N \text{ times}}.$$

In other words,

$$2S = N \cdot (N + 1)$$

or

$$S = \frac{N(N + 1)}{2}.$$

We have learned that the deck of N cards can be put back in order in $N(N + 1)/2 = N^2/2 + N/2$ steps.

What is important here is that the number of steps required is a *polynomial* in N. A problem that can be solved in polynomial time is considered

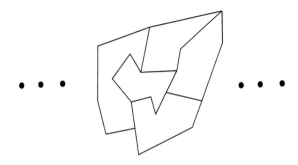

Figure 2.19: Coloring a map with N countries.

by computer scientists to be tractable. If the problem requires exponential time, then the problem is considered to be intractable.

As a simple example of an exponential time problem, suppose that we wish to color a map with N countries as in Figure 2.19. We have four colors to work with: red, blue, yellow, and green. The only rule is that two countries that share an edge cannot have the same color. Is the displayed map four-colorable? Remember that this is not a person solving the problem but a machine. A machine cannot think, but it can try things. In particular, it can try all possible assignments of colors to the N countries and see if one works. Well, the number of things to try is the number of functions from $\{1, 2, \ldots, N\}$ to $\{R, B, Y, G\}$. This is known to be 4^N. In fact you can see this fact easily as follows. The number 1 can be assigned one of 4 values. So it contributes 4 possibilities to the count. The number 2 can be assigned one of 4 values. So it contributes 4 possibilities to the count. And so on. Each of the numbers 1, 2, $\ldots N$ contributes 4 possibilities to the count. So the *total* number of possibilities is

$$4 \cdot 4 \cdot \cdots \cdot 4,$$

where there are N copies of 4 being multiplied together. This is just 4^N.

And that is an exponential function of N. So this problem would be considered to be intractable.

There is copious literature on the topic of computational complexity. The famous P/NP problem—considered by many to be the most important open problem in the mathematical sciences—concerns the question of recognizing problems of polynomial complexity versus problems of exponential complexity. Donald Knuth wrote his article [KNU2] to satirize the situation. We take the opportunity here to describe what Knuth accomplished in his article.

Now Knuth's paper [KNU2] is rather technical, and we shall not reproduce all the details here. Rather our intention is to capture the spirit (and the fun) of what he wrote.

He begins by noting the fairly obvious fact that an arbitrary song of length N (meaning that the song has N words) will require about N storage spaces to remember it. This is bothersome if one needs to learn a great many songs. But some of our ancient ancestors invented the concept of a *refrain*, and that reduces the needed storage space. Note that our understanding, based on years of experience, is that the *verses* can all be different, but the refrain is always the same.

Knuth formulates a formal, mathematical result as follows:

> **Lemma:** Let S be a song of length N (that is, containing N words) containing m verses of length V and a refrain of length R. Here the refrain is to be sung first, last, and between successive verses. Then the amount of storage space required for this song is
> $$\frac{V}{V+R}N\,,$$
> where N is the length of the song.

There is no need to write out the proof of this lemma, because one can see intuitively that a particular fraction of the song consists of the repetitive refrain. And one only needs to store the refrain once.

Knuth concludes this initial discussion by noting that the result can be generalized to a song having verses of different lengths V_1, V_2, ..., V_m provided that the V_j satisfy a certain compatibility condition.

The next major step in Knuth's analysis concerns the Medieval Jewish song *Ehad Mi Yode'a*. This song consists of a refrain R followed by thirteen verses of singing. These are sung in the order

$$V_{13}, V_{12}, V_{11}, \ldots, V_2, V_1$$

followed by

$$V_{12}, V_{11}, V_{10}, \ldots, V_2, V_1$$

followed by

$$V_{11}, V_{10}, V_9, \ldots, V_2, V_1$$

and so forth. Then the refrain is sung again. So we see that 14 units of song lead to $(14^2 + 14)/2$ verses of singing. Stated another way, this says that if N is the length of this song, then it only requires storage space of about $\sqrt{N}$.

This Medieval achievement stood for some time, but was ultimately bested by a Scottish farmer named O. MacDonald. Of course the song is *Old MacDonald Had a Farm*. You probably know the song well enough from your childhood that you can see that this new song entails more repetition than *Ehad Mi Yode'a*, hence has more redundancy, and so stores more song in less space. In fact, instead of $\sqrt{N}$ storage space, Old MacDonald has $\sqrt{N/k}$ storage space for a constant k that is significantly larger than 1.

Knuth notes that the French song *Alouette* and the German song *Ist das nicht ein Schnitzelbank?* achieve similar results.

The examples so far suggest that $\sqrt{N}$ may be a barrier to serious further progress in this problem. But such is not the case. The next advance is credited to the English. According to the legend, the true love of U. Jack gave to him, on successive days, a total of 12 ladies dancing, 22 lords-a-leaping, 30 drummers drumming, 36 pipers piping, 40 maids-a-milking, 42 swans-a-swimming, 42 geese-a-laying, 40 golden rings, 36 collie birds, 30 French hens, 22 turtle doves, and 12 partridges in a pear tree. All this taking place during the twelve days of Christmas. One can calculate that this accounts for $(1/6)m^3 + (1/2)m^2 + (1/3)m$ gifts in m days. So the complexity appears to be on the order of $\sqrt[3]{m}$. But it must be observed that the calculation just presented is based on *the number of gifts* rather than on *the number of units of singing*. A more careful calculation reveals that the complexity is now of order $\sqrt{N/\log N}$. This is important since it breaks the $\sqrt{N}$ barrier.

The logarithm function occurs frequently in mathematics and science. It is the inverse operation to the exponential function. For example, the log base b of x is the power to which we must raise b to obtain x. We write this number as $\log_b x$. As an instance, the log base 2 of 8 ($\log_2 8$) is 3 just because $2^3 = 8$. The log base 4 of 1/4 ($\log_4 1/4$) is -1 because $4^{-1} = 1/4$.

Notice that

$$b^{\log_b x + \log_b y} = b^{\log_b x} \cdot b^{\log_b y} = x \cdot y$$

and

$$b^{\log_b(x \cdot y)} = x \cdot y \,.$$

It follows that

$$\log_b(x \cdot y) = \log_b x + \log_b y \,.$$

This tells us that the logarithm function grows very slowly. For example,

$$\log_b(2x) = \log_b 2 + \log_b x \,.$$

So we see that doubling the argument (from x to $2x$) only increases the value of the function by a constant (namely $\log_b 2$).

For the next big breakthrough, the United States got involved. One W. Blatz of Milwaukee, Wisconsin, discovered a class of songs known as m *Bottles of Beer on the Wall*. After a moment's thought, one realizes that such a song has order of complexity $\log N$.

One can see that $\log N$ complexity is as dramatic an improvement over N complexity as polynomial complexity is over exponential complexity. In some sense, one can think of log N complexity as the final word in the subject. But no ...

Knuth concludes his paper by pointing out that the advent of modern drugs has made it desirable to use even less memory. The ultimate improvement in our estimates is the following song of KC and the Sunshine Band:

> That's the way
> I like it.
> Uh huh, uh huh.
> That's the way
> I like it.
> Uh huh, uh huh.
> That's the way
> I like it.
> Uh huh, uh huh.
> etc.

This song has complexity 1. No matter how long one makes the song, there is only one unit of material to remember. So this is the last word in the subject.

Knuth's last thought is

> It remains an open problem to study the complexity of nondeterministic songs.

Example: The last example, the song of KC and the Sunshine Band, is a bit silly because it repeats the same phrase over and over again. Create an example of a song that has complexity *greater than 1 but arbitrarily close to 1* but does *not* repeat the same phrase over and over.

2.8   Bertrand's Paradox

The subject of probability theory has a rocky history.

A gambler's dispute in 1654 led two famous French mathematicians, Blaise Pascal (1623–1662) and Pierre de Fermat (1601–1665), to lay the basic

Figure 2.20: Dice.

foundations for the idea of probability theory. Antoine Gombaud, Chevalier de Méré (1607–1684), a French nobleman with an interest in gaming and gambling questions, had asked Pascal about an apparent contradiction concerning a popular dice game. [Recall that a die has the numbers 1 through 6, recorded in patterns of dots, on each of its 6 sides. See Figure 2.20.] The game consisted in throwing a pair of dice 24 times, and the problem was to decide whether or not to bet even money on the occurrence of at least one "double six" during 24 throws. A seemingly well-established gambling "folk rule" led de Méré to believe that betting on a double six in 24 throws would be profitable, but his own calculations indicated just the opposite.

This problem and others posed by de Méré led to an exchange of letters between Pascal and Fermat. In these letters, the formal ideas of probability theory were formulated for the first time. Although a few special problems on games of chance had been solved by some Italian mathematicians in the 15th and 16th centuries, no general set of principles was developed before this famous correspondence. Pascal is remembered today for, among other things, Pascal's triangle. Fermat is remembered for many things, but particularly for Fermat's last theorem.

The Dutch scientist Christian Huygens (1629–1695), a teacher of Gottfried Wilhelm von Leibniz (1646–1716), learned of the Pascal/Fermat correspondence and shortly thereafter (in 1657) published the first book on probability; entitled *De Ratiociniis in Ludo Aleae*, it was a treatment of various gambling problems. Because gambling was so popular, probability theory soon became fashionable, and the subject developed rapidly during the 18th century. Key contributors during this period were Jakob Bernoulli (1654–1705) and Abraham de Moivre (1667–1754).

In 1812 Pierre de Laplace (1749–1827) introduced a host of new ideas and mathematical techniques in his book, *Théorie Analytique des Probabilités.* Before Laplace, probability theory was concerned exclusively with studying games of chance. Laplace applied probabilistic ideas to many scientific and practical problems. The theory of error, actuarial (insurance) mathematics, and statistical mechanics are some of the important applications of probability theory developed in the 19th century.

Like so many other branches of mathematics, the development of probability theory has been stimulated by the variety of its applications. Conversely, each advance in the theory has enlarged the scope of its influence. Many scientists have contributed to the theory since Laplace's time; among the most important are Pafnuty Chebyshev (1821–1894), Andrei Markov (1856–1922), Richard von Mises (1883–1953), and Andrey Kolmogorov (1903–1987).

One of the difficulties in developing a theory of probability has been to arrive at a definition of probability that is mathematically precise, yet comprehensive enough to be applicable to a wide range of phenomena. The search took many centuries, and was hobbled by many paradoxes that occurred along the way. The matter was finally resolved in the 20th century by treating probability theory on an axiomatic basis. In 1933 a monograph by a Russian mathematician, A. Kolmogorov, outlined an axiomatic approach that forms the basis for the modern theory. (Kolmogorov's monograph is available in English translation as [KOL].) In this section we treat one of the more famous and baffling paradoxes of probability theory. In particular, we present a problem that has three different correct solutions.

Fix a circle of radius 1. Draw the inscribed equilateral triangle as shown in Figure 2.21. We let ℓ denote the length of a side of this triangle. Suppose that a chord d (with length m) of the circle is chosen "at random." What is the probability that the length m of d exceeds the length ℓ of a side of the inscribed triangle?

The "paradox" is that this problem has three equally valid solutions. We now present these apparently contradictory solutions in sequence. At the end we shall explain why it is possible for a problem like this to have three

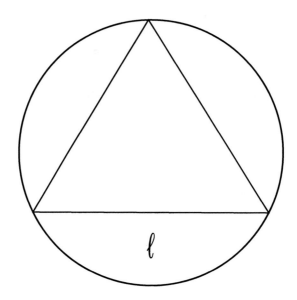

Figure 2.21: The setup for Bertrand's paradox.

distinct solutions.

Solution 1: Examine Figure 2.22. It shows a shaded, open disc whose boundary circle is internally tangent to the inscribed equilateral triangle. If the center of the random chord d lies *inside* that shaded disc, then $m > \ell$. If the center of the random chord d lies *outside* that shaded disc, then $m \leq \ell$. Thus the probability that the length d is greater than the length ℓ is

$$\frac{\text{area of shaded disc}}{\text{area of unit disc}}.$$

But an analysis of the equilateral triangle (Figure 2.23) shows that the shaded disc has radius $1/2$ hence area $\pi r^2 = \pi \cdot (1/2)^2 = \pi/4$. The larger unit disc has area $\pi r^2 = \pi \cdot 1^2 = \pi$. The ratio of these areas is $1/4$. We conclude that the probability that the length of the randomly chosen chord exceeds ℓ is $1/4$.

Solution 2: Examine Figure 2.24. We may as well assume that our randomly chosen chord is horizontal (other positions of the chord may be analyzed in a similar fashion). Notice that if the height, from the base of the triangle, of the chord d is less than or equal to 0 and greater than or equal to $-1/2$, then $m \leq \ell$. Also, if the height is greater than or equal to 1 and less than or

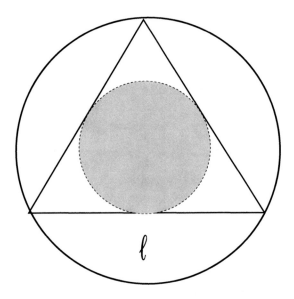

Figure 2.22: The shaded disc.

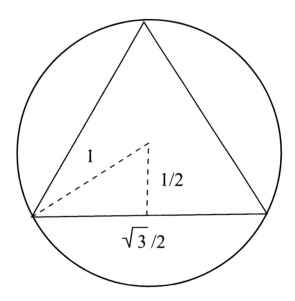

Figure 2.23: Analysis of the equilateral triangle.

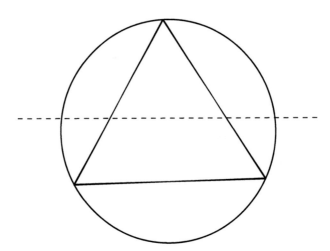

Figure 2.24: A randomly chosen horizontal chord.

equal to $3/2$, then $m \leq \ell$. On the other hand, if the height is between 0 and 1, then $m > \ell$. So the two possibilities occur with equal frequency. We thus see that there is probability $1/2$ that the length m of d exceeds the length ℓ of a side of the equilateral triangle.

Solution 3: Examine Figure 2.25. We may as well assume that one end of our randomly chosen chord occurs at the lower left vertex A of the inscribed triangle. Now look at the angle θ that the chord subtends with the tangent line to the circle at the vertex A (shown in the Figure 2.26). If that angle is between $0°$ and $60°$ inclusive, then the chord is shorter than or equal to ℓ. If the angle is strictly between $60°$ and $120°$, then the chord is longer than ℓ. Finally, if the angle is between $120°$ and $180°$ inclusive, then the chord is shorter than ℓ. In sum, we see that the probability is $60/180 = 1/3$ that the randomly chosen chord has length exceeding ℓ.

How can a perfectly reasonable problem have three distinct solutions probabilities $1/4, 1/3$, and $1/2$? And be assured that each of these solutions is correct! The answer is that, when one is dealing with a probability space having infinitely many elements (that is to say, a problem in which there are infinitely many outcomes—in this case there are infinitely many positions for the random chord), then there are infinitely many different ways to fairly assign probabilities to those different outcomes.

For many years, because of paradoxes such as these, the subject of probability theory was in ill repute. It was not until the invention of a branch of

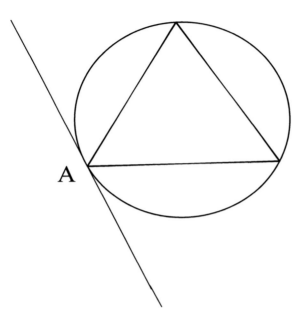

Figure 2.25: The tangent line to the circle at the point A.

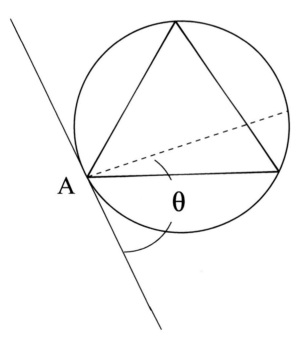

Figure 2.26: The angle that the chord subtends with the tangent line.

mathematics called "measure theory" (Henri Lebesgue, 1906) that the tools became available to put probability theory on a rigorous footing. Problems like these are treated in an advanced course on probability.

Exercise: Come up with a more elementary explanation of why the three different answers provided by Bertrand's paradox are not really contradictory.

Chapter 3

Strategy

3.1 It's All in the Balance

There is an entertaining class of problems that involves a balance scale (Figure 3.1) and a collection of 9 pearls. The premise is that all the pearls but one weigh the same. The odd pearl is either heavier or lighter than the others, but we do not know which. Your job is, in 3 weighings, to find the odd pearl.

Notice that weighing one pearl against one is almost a waste of time. If they do not balance then what have you learned? One of them is the odd pearl, but which one?

If instead you weigh one pearl against one and they balance, then you know that neither is the odd pearl; the odd pearl must be one of the other seven. That gives you two pearls that you can use as "controls," but seven more pearls to check.

Are there possibly other moves to be made with a balance scale? What is to be done? We can make more efficient use of this limited number of options by dividing the 9 pearls into three groups of three. Of course 3 is chosen because it is the only number that evenly divides 9 (other than 9 or 1). We may think of each group of 3 as a "super pearl." Label the groups of three with the names G_1, G_2, and G_3.

Now we weigh G_1 against G_2.

Figure 3.1: The balance scale.

1. If they happen to balance, then all six pearls in G_1 and G_2 are control pearls. The odd pearl is one of the pearls in G_3.

2. If they do not balance, then all the pearls in G_3 are control pearls. The odd pearl is either in G_1 or G_2 but we do not know which.

We first consider Case 1. For Step 2, weigh G_1 against G_3. Of course they will not balance. G_3 will either be lighter or heavier. Make a note of which. Say that, for the sake of specificity, G_3 is heavier. That means that the odd pearl is heavier than the others, and it lies in G_3. Now select any two pearls from G_3 and weigh them against each other. If they balance, then the odd pearl is the third pearl in G_3 and it is heavier. If they do not balance, then the heavier of the two is the odd pearl.

Now consider Case 2. Make a note of whether G_1 or G_2 is heavier. For specificity, say that G_1 is heavier. Now weigh G_1 against G_3. If they balance, then the odd pearl is in G_2 and it is lighter. Pick any two pearls from G_2 and weigh them against each other. If they balance, then the odd pearl is the third one from G_2 and it is lighter. If they do not balance, then the lighter of the two is the odd pearl.

If instead G_1 and G_3 do not balance, then the only possibility is that G_1 is heavier than G_3 (otherwise there would be three weight categories, which is impossible). So the odd pearl is in G_1, and it is heavier. For the last step, weigh any two pearls from G_1 against each other and proceed as in the earlier cases.

Notice that, once we realized that our best strategy was to divide into groups of three and create three "super pearls," then our succeeding steps were almost automatic. If we had instead divided into groups of $\{2, 2, 5\}$ or $\{4, 4, 1\}$, then we would not have known what to do after Step 1.

Suppose now that you have 12 pearls, all appearing the same but with one having an odd weight. You do not know whether the odd pearl is heavier or lighter. How many weighings are needed to find the odd pearl?

The solution that we gave to the previous problem seems to be "tight." And it *is*, insofar as use of the ideas that we have considered so far is concerned. Therefore, if we are going to succeed in finding the odd pearl among 12—using only three weighings—then a new idea will be required.

Begin by using the "super pearl" idea. Divide the twelve pearls into three groups of 4. Call them G_1, G_2, G_3. As a first step, weigh G_1 against G_2.

1. If they happen to balance, then all eight pearls in G_1 and G_2 are control pearls. The odd pearl is one of the pearls in G_3.

2. If they do not balance, then all the pearls in G_3 are control pearls. The odd pearl is either in G_1 or G_2 but we do not know which.

We first consider Case 1 (which is relatively easy). Take any three pearls from G_1 and weigh them against any three pearls from G_3. If they balance, then the odd pearl is the remaining pearl from G_3. Weighing that last pearl against one of the pearls from G_1 will tell whether the odd pearl is heavy or light. If they do not balance, then the odd pearl will be among the three selected from G_3 and we will know whether it is lighter or heavier (since the pearls from G_1 are control pearls). Now a third weighing, as usual, will pin down the odd pearl from among those three that we selected from G_3.

For Case 2, we suppose for specificity that G_1 is heavier and G_2 lighter. Give the pearls in G_1 the names a, b, c, d and give the pearls in G_2 the names a', b', c', d'. For the second weighing, we weigh $\{a, b, a'\}$ against $\{c, d, b'\}$.

(a) If they balance, then the odd pearl is one of c', d' (the two pearls from G_1 and G_2 that we omitted from this second trial). Of course c', d' come from the light side, so we know that the odd pearl is light. For the third weighing, we weigh c' against d'. The odd pearl is the lighter of the two.

(b) If they do not balance, then say that $\{a, b, a'\}$ is heavier. This must mean that c, d are control pearls, and so is a', or else the balance would be the other way. Thus the odd pearl is either $a, b,$ or b'. Finally, weigh a against b. If they balance, then the odd pearl is b' and it is light. If they do not balance, then the odd pearl is the heavier of the two (since a and b come from G_1).

(c) The case that $\{c, d, b'\}$ is heavier is handled just as in subcase (b).

Exercise: We shall end this section with a mathematical joke. Let us show that you can find the odd pearl from among *any number of pearls* with just 3 weighings. For instance, if you have 10 billion pearls, and one of them weighs differently (either heavier or lighter), then you can find it in 3 weighings.

The proof is by induction. Clearly you can handle 1 or 2 or 3 pearls with just 3 weighings (in fact, in the discussion above we handled 9 pearls and 12 pearls). So assume that we know how to handle k pearls. Now you are faced with $k+1$ pearls. What do you do? You put one pearl in your pocket. What remains is k pearls, and you simply run your k-pearl algorithm on those k pearls. If you find the odd pearl (as you should), then that is indeed the odd pearl and you are done. If you do *not* find the odd pearl then it must be the one in your pocket!

End of story. What is the error in the reasoning of the last paragraph?

3.2 See and Say

The following sequence has become known as the "See and Say Sequence," so named by noted Princeton mathematician John Horton Conway. Can you determine the next term of the sequence?[1]

$$1, 1, 1, 3, 1, 4, 1, 1, 3, 6, 1, 2, 3, 1, 4, 8, 1, 3, 3, 2, 4, 1, 6 \ ?$$

---

[1]This type of annoying problem frequently occurs on SAT tests and other intelligence tests. We promise you that the discussion here will be *much more entertaining* than any SAT test that you ever took.

Conway was told of this sequence by one of his students. Now he was one of the best in the world at solving problems of this kind, and he tormented himself for two weeks trying to find the next term of the sequence. He could not do it, and ultimately had to throw himself on the student's mercy and ask the answer. Here it is.

You begin the sequence thusly:

$$1$$

Then you read out loud what you see: "one one." And those are the next two terms of the sequence:

$$1, 1, 1$$

Now you read out loud what you see: "three ones." And those are the next two terms of the sequence:

$$1, 1, 1, 3, 1$$

Next you read out loud what you see: "four ones, one three." And those are the next four terms of the sequence:

$$1, 1, 1, 3, 1, 4, 1, 1, 3$$

You continue to generate terms of the sequence by seeing and saying. Hence the somewhat eponymous name for this sequence.

The hallmark of a great mathematician is that he/she knows how to snatch victory from the jaws of defeat. And that is just what John Conway did in this situation.

Conway's Cosmological Theorem: Every "see and say" sequence eventually splits ("decays") into a sequence of "atomic elements," which are finite subsequences that never again interact with their neighbors. There are 92 elements containing the digits 1, 2, and 3 only, which John Conway named after the natural chemical elements. There are also two "transuranic" elements for each digit other than 1, 2, and 3.

The terms eventually grow in length by about 30% per generation. In particular, if L_n denotes the number of digits of the nth member of the sequence, then the limit of the ratio L_{n+1}/L_n exists and is given by

$$\lim_{n \to \infty} \frac{L_{n+1}}{L_n} = \lambda$$

where $\lambda = 1.303577269034\ldots$ is an algebraic number of degree 71. This fact was proven by Conway, and the constant λ is known as *Conway's constant*.

Exercise: Try to understand Conway's Cosmological Theorem. Can you identify the atomic elements? Can you verify that they do not interact with their neighbors?

3.3   The Ponzi Scheme

The most famous con game of all is the "Ponzi scheme," named after Charles Ponzi (1877–1949). Although Ponzi-like schemes were described in some of Charles Dickens's novels (such as *Martin Chuzzlewit*), it was Ponzi who put them into action in real life. Especially in the 1920s he implemented some daring schemes. He ultimately went to jail, and even worked some Ponzi schemes while he was incarcerated.

A typical Ponzi scheme works like this. Say that a very desirable computer costs \$10,000. You approach your "mark" (i.e., your victim) and tell him that you can obtain this computer for him/her for a mere \$5,000. The only catch is that they must give you the money up front, and they must be willing to wait 2 months for delivery.

So far so good. What the mark doesn't know is (i) that you are paying the full price of \$10,000 for each computer but (ii) that you are taking orders at triple the rate at which you are making deliveries.

Thus, in the first month, let us say that you take ten orders but deliver no computers. In the second month you take twenty more orders but only deliver computers to your ten customers from the first month. We see that by the end of the second month you have taken in $30 \times \$5,000$ but have only delivered ten computers at \$10,000 each. You have made a profit of \$50,000.

Clearly, if the con artist is to make money, he must at some point disappear, leaving a (large) number of orders unfilled. Let us design a Ponzi scheme so that, at the end of one year, the con man can make away with an even one million dollars.

We shall imagine that the con artist is marketing a Mercedes Benz that usually sells for \$50,000. But he tells the mark that he can provide a Benz for

$25,000 provided that (a) the mark is willing to pay the full price up front, (b) the mark is willing to wait 2 months for delivery, and (c) the mark agrees that he/she will receive no paperwork on the car. Let us suppose that the con artist sells X number of cars per month but only delivers on $X/2$ of the orders per month. Then the con is pocketing

$$25000 \cdot (X/2) \cdot 12$$

dollars in a year. The factor of 12 is in there because there are 12 months in a year. So, in sum, the con man is making $150000 \cdot X$ dollars per year. Clearly, if $X = 7$, then he will make his million.

The largest financial fraud in United States history was that perpetrated by Bernie Madoff. He was caught in 2008, but the scheme may have been going on for as long as 30 years. And it was a massive Ponzi scheme that ultimately netted Madoff a total of $18 billion dollars. Many high-flying investors were taken by Madoff's Ponzi scheme, and a number of them lost a substantial portion of their assets. Madoff got a 150-year jail sentence for his wrongdoings.

Exercise: Read about Bernie Madoff in `Wikipedia` or some other online source. Can you figure out what his Ponzi scheme was, and how it worked? Why were so many people taken in by Madoff? What precautions can *you* take so that you are not swindled in this way?

3.4 Ham Sandwich Theorems

It is perhaps surprising to think that mathematicians have spent their time ruminating about ham sandwiches, but they have. These ideas grew out of the rapid and dynamic development of topology in the early twentieth century.

Here is a version of the question. Suppose that, in the plane, you have a set consisting of ham and a set consisting of bread. These two sets are, of course, disjoint. Apart from that, they can be any old shape. See Figure 3.2.

Is there a single line that cuts each of the sets in half? That is to say, is there a single line that divides the area of the ham in two and divides the

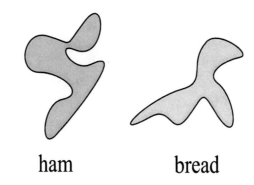

ham   bread

Figure 3.2: A "ham sandwich."

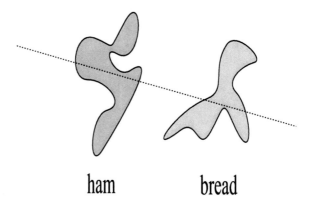

ham   bread

Figure 3.3: A single cut that bisects the ham and bisects the bread.

area of the bread in two? See Figure 3.3 for a suggestion of how this can be done with the ham and bread in Figure 3.2.

This is really quite startling. Why should we believe that such a line exists? Here is an intuitive argument to help you to believe the result.

Imagine a large circle in the plane, centered at the origin, that contains both the ham and the bread. See Figure 3.4.

Let P be a point on that circle. The vector from the origin 0 out to P gives a direction, and there is a set of lines that is perpendicular to that direction. See Figure 3.5.

One of those lines that is perpendicular to the indicated vector will bisect the bread and one of those lines will bisect the ham. See Figure 3.6.

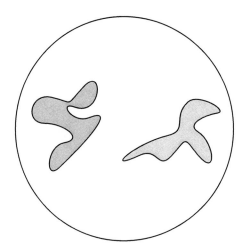

Figure 3.4: A circle that contains the ham and the bread.

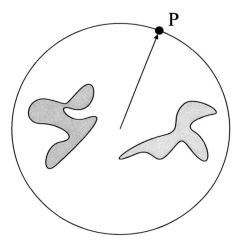

Figure 3.5: The vector out to P.

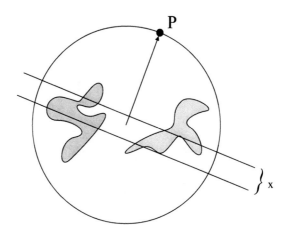

Figure 3.6: A line that bisects the ham and a line that bisects the bread.

Let x be the distance between the two lines. As P moves around the circle, x varies continuously. But, for some value of P, the two lines will switch position and we will be forced to allow x to take negative values. But that means that, for some value of P, x takes the value 0. What does that mean? It means that, for that particular direction, the line that bisects the bread is zero distance from the line that bisects the ham. Which means that they are the same line. In other words, we have found a single line that bisects both the ham and the bread. This is the ham sandwich theorem.

Naturally mathematicians, once they have established a result, like to generalize it. So we might ask what is true in three dimensions? And what is true in even higher dimensions? This new set of questions has become known as the *generalized ham sandwich theorem*.

It was Hugo Steinhaus who posed the problem in three dimensions, and Stefan Banach who solved it. Later, in 1942, Stone and Tukey proved the result in all dimensions.

In three-dimensional space, a generalized ham sandwich consists of some bread, some ham, and some cheese. Note that, as before, the ham can be a blob of any old shape and the cheese can be a blob of any old shape and the bread can be a blob of any old shape. See Figure 3.7.

The question now is whether there is a single plane that bisects the ham, bisects the cheese, and bisects the bread. And the answer is yes! What about in higher dimensions? In n dimensional space, we can have n disjoint sets.

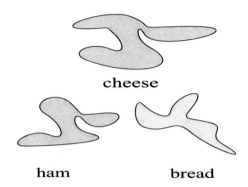

Figure 3.7: A generalized ham sandwich in three space.

And the result is that there is an $(n-1)$ dimensional hyperplane that bisects each of the n sets. There is your generalized ham sandwich theorem.

Example: An historically famous theorem, too sophisticated to treat in any detail here, but related to the ham sandwich theorems, is the Borsuk-Ulam theorem. This result says that any continuous, scalar-valued function on the sphere takes the same value at two antipodal points. Here two points are antipodal if they are at opposite ends of a diameter of the sphere. A nice example of Borsuk-Ulam is the temperature function on the surface of the Earth. We know heuristically that this function is continuous. So we may conclude that there are two antipodal points on the surface of the Earth at which the temperature is the same.

Give an example of a continuous temperature function on the sphere $x^2 + y^2 + z^2 = 1$ and find the two antipodal points at which the function takes the same value.

Chapter 4

Focus

4.1 The Erdős Number

The Hungarian scholar Paul Erdős (1913–1996) was one of the most remarkable mathematicians of the twentieth century. He never had a regular job and he never had a regular home. He just traveled from university to university, working with various mathematicians on a bewildering variety of problems. In the end, Erdős wrote over 1500 mathematical papers, several books, and had over 500 collaborators (this author was one of them).

According to Casper Goffman (1913–   ) [GOF], your Erdős number is zero if you are Paul Erdős. Your Erdős number is $\leq n$ if you have written a paper with somebody whose Erdős number is $n - 1$. Put in other words:

- Your Erdős number is 0 if you are Paul Erdős.

- Your Erdős number is 1 if you have written a paper with Paul Erdős.

- Your Erdős number is 2 if you have written a paper with someone whose Erdős number is 1.

- And so forth.

Goffman's *Monthly* article was answered by Erdős [ERD], who expressed his delight with the concept and asked the following question: If each mathematician is a node on the face of the Earth, and if any two mathematicians

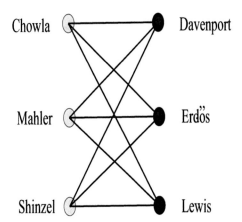

Figure 4.1: A non-embeddable complete bipartite graph on 6 vertices.

who have collaborated are connected with an edge, then is the resulting graph $\mathcal{G}$ planar? The answer "no" is also provided by Erdős: the graph $\mathcal{G}$ contains a copy of $\mathcal{R}(3,3)$, the complete bipartite graph on 6 vertices. That is, there are three vertices of each color black and white, and there are nine edges connecting black to white in all possible ways. Such a graph is non-embeddable. The whites in this case are S. Chowla (1907–1995), Kurt Mahler, and Andrzej Schinzel. The blacks are H. Davenport, Erdős, and D. J. Lewis. Refer to Figure 4.1.

I also hosted Erdős at Penn State, where he gave two lectures. One day we took him to lunch. He ordered a fruit smoothie to drink, and it was served (as is customary) with a straw inserted and the straw paper left over the exposed end of the straw. Erdős made a huge show of trying to drink the beverage through the paper. At the end of the meal, the waitress asked us if we wanted coffee. Everyone said "yes" except Erdős. He said, "I have something much better than coffee," and then he made a big production of bringing out his benzedrine.

Just so, there is now a concept of "Bacon number." What could this be? Kevin Bacon (1958–) is a hard-working and prolific movie actor who seems to have made a movie with practically everyone. So your Bacon number is 1 if you have made a movie with Kevin Bacon. Your Bacon number is 2 if you have made a movie with someone who has Bacon number 1. And so forth.

It is quite difficult to find anyone with Bacon number greater than 3. The Web site http://www.cs.virginia.edu/oracle/ will assist you in your search. You will find that Jack Nicholson has Bacon number 1. Laurence Olivier has Bacon number 2. So do James Cagney and Drew Barrymore. Douglas Fairbanks has Bacon number 3. So does Mary Pickford (arguably the very first movie star). Interestingly, Paul Erdős has Bacon number 4. Is it possible? The provenance is this:

Paul Erdős was in *N is a Number* (1993) with Gene Patterson;

Gene Patterson was in *Box of Moon Light* (1996) with John Turturro;

John Turturro was in *Cradle Will Rock* (1999) with Tim Robbins;

Tim Robbins was in *Mystic River* (2003) with Kevin Bacon.

The "average Bacon number" is about 2.941. There are 7452 actors with Bacon number 5, 944 with Bacon number 6, 107 with Bacon number 7, and 13 with Bacon number 8. But it turns out that the more senior actor Sean Connery has a better record. The "average Sean Connery number" is about 2.706. It turns out, however, that actor Rod Steiger is the best "center" of the Hollywood universe. His average is about 2.6519.

As previously noted, this author has Erdős number 1.

Exercise: The American Mathematical Society utility `MathSciNet` has a feature that will calculate the Erdős number of any mathematician. If your college has access to `MathSciNet`, then learn how to use this feature and calculate the Erdős number of some of your math instructors.

4.2 Time Out

Everyone knows that a classical analogue clock has both hands coincide at noon. See Figure 3.9. What is the next time after that when both hands coincide?

It will obviously be some time after 1:00 p.m. At 1:00 p.m., the big hand is to the left of the little hand. At 1:30 p.m. the positions are reversed. So—at some time in between, the hands must coincide.

Figure 4.2: A classic analogue clock.

Let L denote the position of the big hand measured in minutes with $0 \leq L < 60$, and let l denote the position of the short hand measured in hours with $0 \leq l < 12$, (i.e., at three o'clock, $L = 0$ and $l = 3$). Then we must have that l and L satisfy the equation

$$k + \frac{L}{60} = l\,,$$

where $k = 0, 1, 2, \ldots, 11$. The hands meet when $5l = L$. Thus we have that, in this case, l satisfies

$$k + \frac{l}{12} = l.$$

Thus the hands will meet at $l = \frac{12k}{11}$ hours and $L = \frac{60k}{11}$ minutes, with $k = 0, 1, 2, \ldots, 11$. Now, since for $k = 11$ we obtain $l = 12$ and $L = 60$ (which actually corresponds to $l = 0$, $L = 0$, which is the case $k = 0$), we have that the hands meet 11 times in a 12-hour period.

In particular, for $k = 1$ we get $l = 12/11 \approx 1.0909$ and $L = 60/11 \approx 5.4545$. So the first time after noon when the hands coincide is at 5.4545 minutes after 1:00 p.m.

Exercise: Calculate the next time (after the one that we calculated in this section) at which the two hands of the clock coincide. This will be some time after 2:00 p.m.

Can you calculate the first time after noon that the two hands are diametrically opposite?

4.3 Days of the Week

On which of Saturday or Sunday does New Year's Day fall most frequently?

In the Gregorian Calendar, there are 52 weeks and 1 day in a normal year. Leap years have 1 extra day, February 29th. Hence, if year x starts on a Saturday, year $x + 1$ will start on a Sunday if x is a normal year; otherwise it will start on Monday.

Recall that all years that are divisible by 4 are leap years, except the years that are divisible by 100, in which case they are leap years only if they are divisible by 400. So year 2100 is not a leap year, whereas year 2000 is.

In view of the above remarks, the years x and $x + 400$ start with the same day of the week, no matter which year x is. This implies that to know the relative frequencies by which New Year's Day falls on different days of the week, it is enough to know these frequencies only in a period of 400 years.

It would be a terribly boring and tedious job if we were to sit down and write 400 consecutive years and their corresponding New Year's Days. Fortunately, there are steps that we can take to make the job easier.

First it is obvious that each period of 400 years has one and only one year that is divisible by 400. If it were not for this year, we could divide the interval of 400 years into 4 subintervals of 100 years, and then find how many times each particular day becomes New Year's Day in the first 100 years. Then, since all the other 3 subintervals would have had the same pattern, except that each starts on a different day, by renaming the days of the week we could find the number of times each day becomes New Year's Day in the other subintervals. Finally, by adding the figures we could answer the question.

We can do this if we choose the period so that the "troubling" year is at the end of one of the subintervals, preferably at the end of the interval itself. For example, the interval from 2001 to 2400 will do.

Before starting to count, let us refer to each day of the week by numbers $1, 2, \ldots, 7$, assuming that the year 2001 starts with day 1. We will give the appropriate names to these days at the end.

To make counting even easier, we notice that in each interval of 28 years that does not include the years that are divisible by 100, New Year's Day falls on each day of the week exactly 4 times. We leave it to the reader to either prove this or convince himself/herself by counting. Thus, by the year 2084 each day will have been New Year's Day exactly $12 = 3 \times 4$ times. We only have to include the last sixteen years which we do simply by counting. Knowing that the years 2001 and 2085 start on the same day we have

$$1 , 2, 3 , 4 , 6 , 7 , 1 , 2 , 4 , 5 , 6 , 7 , 2 , 3 , 4 , 5$$

as New Year's Days for the last 16 years. Notice that the next period of one-hundred years will start with day 6.

We summarize all the results in the following table.

n	1	2	3	4	5	6	7
$A_1(n)$	14	15	14	15	14	14	14

Here $A_i(n)$ denotes the number of times New Year's Day falls on day n during the ith subinterval.

As we mentioned before, the next hundred years starts with day 6. Hence, by relabeling the above table, we get

n	6	7	1	2	3	4	5
$A_2(n)$	14	15	14	15	14	14	14

.

Similarly we get the following tables.

n	4	5	6	7	1	2	3
$A_3(n)$	14	15	14	15	14	14	14

n	2	3	4	5	6	7	1
$A_4(n)$	14	15	14	15	14	14	14

Adding the corresponding figures for each day we get Table 4.1, in which $A(n)$ refers to the number of times day n will be New Year's Day in the period 2001 to 2400. Now it only remains to give appropriate names to each of the numbers.

Table 4.1

n	1	2	3	4	5	6	7
$A(n)$	56	58	57	57	58	56	58

This can be done easily by checking a calendar to see that the year 2001 starts on a Monday, so in Table 4.1, Saturday is day 6, and Sunday is day 7. Thus, New Year's Day falls more often on Sunday than on Saturday.

Exercise: Can you determine on which day Christmas falls most frequently?

Chapter 5

Science

5.1 A Belt for the Earth

The circumference of the Earth is about 25,000 miles. Suppose that you have a steel band that encircles the Earth along the Equator, and is clamped tightly to the Earth. So this band is 25,000 miles long. See Figure 5.1.

By how much would you have to lengthen the band so that it stands 1 foot above the Earth at all points (see Figure 5.2)?

Your intuition may tell you that this new band will have to be quite a lot longer. This is one situation in which your intuition is *way off*.

In order to answer this question, we must remember that the circumference of a circle of radius r is $2\pi r$. See Figure 5.3.

We know that the original band has length (remember that a mile equals 5280 feet)

$$C = 25,000 \text{ miles} = 132,000,000 \text{ feet}.$$

Therefore the radius of the circle that this band describes is

$$r = \frac{132,000,000}{2\pi} = \frac{132,000,000}{2 \cdot 3.14159} = \frac{132,000,000}{6.28318} \approx 21,008,470.23 \text{ feet}.$$

The new band, which stands 1 foot off the surface of the Earth, will have radius

$$r' = r + 1 = 21,008,470.23 + 1 = 21,008,471.23.$$

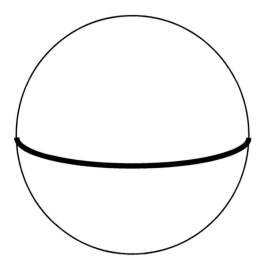

Figure 5.1: A band clamped around the Equator.

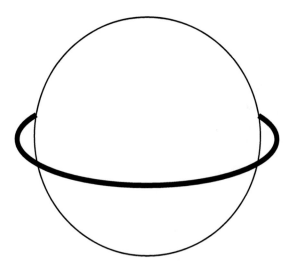

Figure 5.2: The enlarged band that stands 1 foot off the Earth's surface.

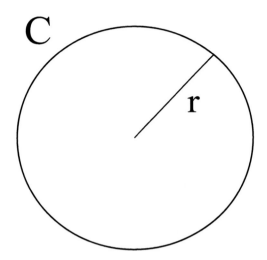

Figure 5.3: The circumference of a circle.

Thus the new circumference will be

$$C' = 2\pi \cdot 21,008,471.23 \approx 6.28318 \cdot 21,008,471.23 = 132,000,006.3 \text{ feet}.$$

We see, in conclusion, that the new band exceeds the old band in length by

$$C' - C \text{ feet} = (132,000,006.3 - 132,000,000) \text{ feet} = 6.3 \text{ feet}.$$

The surprising result of our calculation is that, in order to expand our tight band around the Equator to one that stands one foot off the surface of the Earth, we must lengthen the band by 6.3 feet.

You may want to think about the fact that the Earth has nothing to do with the result of this problem. The same calculation performed for Jupiter or Saturn or Pluto will still yield that it requires a lengthening of 6.3 feet in order to make the band stand 1 foot off the surface of the planet.

Exercise: Suppose, just for the fun of it, that the Earth were cubical in shape. Then how would your answer to this question change?

5.2   Your Next Breath

Sir James Jeans was a distinguished physicist who was famous for, among other things, posing the following question: What is the likelihood that the next breath you take will contain a molecule of air that was expelled by Julius Caesar when he exclaimed, "Et tu Brute?"

The naive answer to this question is simply laughter. It sounds virtually impossible. But it is not, and one can do a fairly straightforward physical calculation to answer the question. For further reading on this matter, see [LIT], [PAUL1], and [REN].

First, we cannot know everything. We must make certain assumptions. One is that the last breath that Julius Cæsar exhaled is uniformly distributed throughout the atmosphere. A second is that all the molecules that were in that breath are still *present* in the atmosphere—they haven't dispersed to parts unknown in the universe, and they haven't decomposed and recombined with other elements (for instance, in a process of oxidation). Finally, we shall assume that molecules of air are evenly distributed in the atmosphere (this is not strictly true, as the atmosphere becomes more rarefied the further we go from the surface of the Earth—however near the surface of the Earth, where we actually live, the hypothesis is approximately true).

We also need some concrete physical data, which we will borrow from a standard reference. First, one may use the *CRC Handbook of Chemistry and Physics* [CRCCP] to find the mass of our atmosphere, the magnitude of Avogadro's number, and the gram molecular weight of the atmosphere. The result is that the atmosphere contains 10^{44} molecules.

The gram molecular weight of any gas at standard temperature fills 22.4 liters and contains 6×10^{23} molecules. Experimentation shows that an average breath contains 0.4 liters of air. Thus the number of molecules in an average breath is

$$0.4 \cdot \frac{1}{22.4} \cdot \left[6 \cdot 10^{23}\right].$$

This amounts to 1.0714×10^{22} molecules.

Thus the situation is quite simple: your next breath has 1.0714×10^{22} molecules, Cæsar's last breath had 1.0714×10^{22} molecules, and these Cæsarian molecules are mixing around in a universe of 10^{44} molecules. What is the probability that the two breaths will overlap (i.e., have at least one molecule in common)?

First we discuss the matter intuitively. Round off the number of molecules per breath to 10^{22}. Since the atmosphere has 10^{44} molecules, that makes 10^{22} total breaths in the atmosphere, each having 10^{22} molecules. If Cæsar's last breath of 10^{22} molecules is equally and randomly distributed in the atmosphere, then there is likely one molecule of that last breath in each of the other breaths (for there is one of Cæsar's molecules for each of the breaths in the atmosphere). So it seems nearly certain that your next breath will contain one of Cæsar's molecules.

The trick is to make this precise. And the process of making the calculation precise introduces us to a major problem with scientific computing. According to the preceding simplified figures, the atmosphere contains $10^{44} - 10^{22}$ molecules that are not Cæsar's. Pick a molecule of your next breath. The probability that that is a non-Cæsarian molecule is

$$\frac{10^{44} - 10^{22}}{10^{44}} = 1 - 10^{-22}. \tag{5.2.1}$$

This probability applies to each molecule in your next breath. So the probability that *every* molecule in your next breath is non-Cæsarian is the product of the number in line (5.2.1) with itself 10^{22} times (once for each molecule in your next breath). Thus the probability that your next breath is entirely non-Cæsarian is

$$= \left(1 - 10^{-22}\right)^{10^{22}}. \tag{5.2.2}$$

Now here is the rub: if you punch in the number $1 - 10^{-22}$ on your calculator, you will just get 1—because your calculator is probably only accurate to 10 digits at most. The number in (5.2.2) is most assuredly *not* one. But how can we get our hands on it? Most out-of-the-box computer languages (such as FORTRAN) are accurate to eight digits—sixteen digits in double precision. How can we proceed?

Calculators and computers are no substitute for theoretical mathematics, and theoretical mathematics is virtually the only tool that can save us now (you might, however, try the computer algebra package Mathematica if it is available to you). It is known (check a book of tables such as [CRC]) that the expression $(1 - 1/k)^k$ tends to Euler's number $e \approx 2.718\ldots$ as $k \to \infty$. It is further known that $(1 - 1/k)^k$ approximates e to an accuracy of k decimal places. Our number $(1 - 10^{-22})^{10^{22}}$ fits this model exactly with $k = 10^{22}$. We conclude that the probability of our next breath *not* containing a molecule of Cæsar's last breath is

$$\left(1 - 10^{-22}\right)^{10^{22}} \approx \frac{1}{e} \approx \frac{1}{2.718} = 0.368.$$

In other words, the odds are at least 63% that your next breath *will* contain a molecule of Julius Cæsar's last breath.

Exercise: What are the odds that your next breath will contain *two molecule* of Julius Cæsar's last breath?

5.3   A Hairy Question

How fast does hair grow in miles per hour? The first reaction of many people to this question is ridicule: Hair doesn't grow in miles per hour! What are you talking about?

Well, this thinking is incorrect. In most accounts that one can read on the Internet, the growth of hair is described in terms of inches per month. But inches can be converted to miles, and months can be converted to hours. So the question can, at least in principle (and with the aid of a calculator) be answered. The Web site

https://en.wikipedia.org/wiki/Human_hair_growth

will tell you that human hair typically grows at a rate of 0.5 inches per month. Now a mile is 5280 feet, or 63,360 inches. So hair grows at a rate of $0.5 \cdot (1/63360) = 0.000007891$ miles per month.

Well, a month is 30 days and a day is 24 hours. So a month is 720 hours. Thus hair grows at the rate of 0.000007891 miles per 720 hours, or 0.000000011 miles per hour.

Exercise: Calculate how fast the height of a baby increases in units of light years per century.

5.4 The Motions of the Planets

Man has been fascinated with the motions of the planets even since ancient times. The ancients were particularly noted for attempts to calculate and anticipate lunar and solar eclipses. But it was only in the Renaissance that people took an analytical approach to the matter. The main exponent was Johannes Kepler (1571–1630). Kepler's story is a fascinating one, and we recount it here.

Kepler spent his entire adult life studying the motions of the planets. He wanted to formulate laws describing these motions, but he needed data to evaluate and study in order to determine what those laws might be. His teacher Tycho Brahe (1546–1601), who was a well-established and distinguished astronomer, had that data. But he would not release it to Kepler. Why not?

There were two prevalent theories of the structure of the solar system in those days. The traditional Ptolemaic (or Biblical) view was that the Earth is at the center of the solar system and the other solar system bodies orbit around it. The more modern Copernican heliocentric theory is that the sun is at the center of the solar system and the planets orbit around *it*.

Now Brahe was a partisan of the Ptolemaic theory. He was afraid that, were he to give his copious data to Kepler, then Kepler would use it to develop and promote the Copernican theory. Both men were hard-headed, dedicated scholars, disinclined to give an inch. So matters were stuck at a standstill.

But fate intervened. In the days of Tycho Brahe and Kepler, academic positions were scarce, and there was great rivalry and jealousy among scholars. If one was lucky, one could get a wealthy sponsor, and then he/she did not need to worry about an academic position. Brahe was lucky in that regard. He got a sponsor who bought an island for him and set up an observatory so that Brahe could do his astronomical work. It was an ideal setup.

The only tithe that Tycho Brahe had to pay was that, on occasion, his sponsor would put on grand balls and parties. And he required Tycho Brahe to attend so that he could show the scholar off to his cronies. At one such get-together, Tycho Brahe drank too much beer; his bladder burst and he died.

This was Kepler's good fortune, for he was able to negotiate with Brahe's family and obtain the much-longed-for astronomical data that Tycho Brahe had amassed over the years. Kepler in turn spent many years studying this data, and he came to certain profound conclusions. It should be understood that what Kepler had was thousands of pages of numerical recordings that Tycho Brahe had performed at his telescope. Of course there were no computers in those days, and Kepler had to do all his calculations by hand. It so happened that John Napier (1550–1617) had developed his theory of the logarithm at exactly that time, and logarithms would certainly have streamlined Kepler's task by an order of magnitude. But Kepler could not understand the theory underpinning the idea of logarithms, so he refused to use them. He did all his calculations in the most primitive, labor-intensive fashion. This took many years.

But the end result was startling, elegant, and fascinating. Kepler ended up formulating three laws of planetary motion. What is striking about these laws is

- they are easy to understand and to use,

- they are correct,

- they still stand today as basic to our understanding of planetary motion.

And it should be stressed that Kepler did not derive these laws from any analysis or physical reasoning. Instead, he stared at reams of numbers, he calculated, and he slowly derived certain physical conclusions.

Kepler's Laws are these:

1. The orbit of each planet is an ellipse, with the sun at one focus. See Figure 5.4.

2. The line segment from the center of the sun to the center of an orbiting planet sweeps out area at a constant rate. See Figure 5.5.

3. The square of the period of a planet's revolution is proportional to the cube of the length of the major axis of its elliptical orbit, with the same constant of proportionality for any planet. See Figure 5.6.

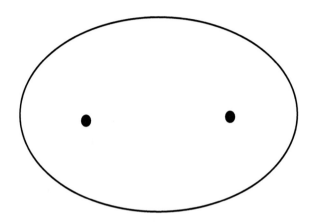

Figure 5.4: The orbit of each planet is an ellipse.

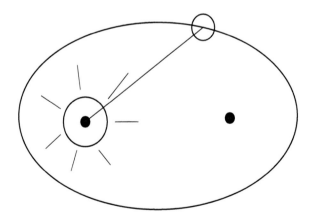

Figure 5.5: Area is swept out at a constant rate.

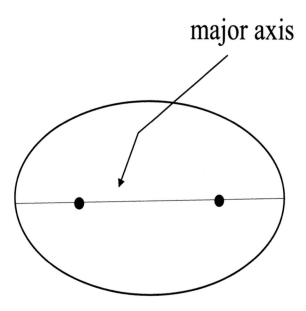

Figure 5.6: The square of the period is proportional to the cube of the major axis.

And now the story gets even more interesting. The great scientist Isaac Newton (1643–1727) was an eccentric and difficult individual. He had few friends. One of his friends was Edmund Halley of Halley's comet (1656–1742). One day Newton had a visit from Halley, and Halley related to Newton that he and a group of scientists had spent twenty years trying to decide whether Kepler's Laws of planetary motion could be derived analytically from first principles—in particular from Newton's inverse square law of gravitation. Without hesitation Newton said, "Yes, they can. And I know how to do it."

Well, you can imagine that Halley reacted rather strongly. He immediately told Newton that this was a result of the most immense importance, and demanded the details. Newton said that he thought that result was not of any particular interest, and he had discarded his notes. Well, then Halley really went into a tizzy. He importuned Newton to write these results up carefully and to share them with the world. It should be noted here that Newton's career is marked by many incidents of his not publishing his work in a timely manner. The great dispute over whether Newton or Leibniz invented calculus was sparked in part by Newton's reticence to publish.

Newton finally acquiesced to Halley's demands, and as a result he ended up writing his magnum opus entitled *Principia Mathematica*. This book is commonly thought to be the greatest scientific work ever written. It explains Newton's foundational ideas about calculus, about optics, and many of his other deep insights. It explains the inverse square law of gravitation, and it derives Kepler's Laws from first principles.

If you take a course in multi-variable calculus, then you will learn about Newton's derivation of Kepler's Laws from first physical principles. This was actually one of the very first modern physical analyses in history. It is groundbreaking work that has influenced science now for more than three hundred years. We cannot present the details of Newton's analysis here, but we can celebrate his achievement.

Exercise: Use Kepler's third law, together with your knowledge of the length of a year on the Earth, to determine the major axis of the Earth's orbit.

5.5 How Big Is Big Data?

What is statistics? It is the science of trying to get information from a collection of numbers. Just as an instance, imagine giving an exam to 25 students. The exam is worth 100 points. Suppose that the scores on this exam are as follows:

$$47, 51, 52, 52, 55, 55, 55, 60, 63, 63, 69, 69, 69,$$

$$74, 82, 82, 86, 89, 93, 93, 95, 97, 97, 100, 100 \, .$$

How can we derive a single number that will tell us something about the performance of the class taken as a whole?

The single simplest idea in the subject is to take the *average* or *mean*. You add together all these scores and divide by 25. The answer is $\mu = 73.92$. This is a nice number, because 13 students scored less than μ and 12 students scored more than μ. So the average μ is right in the middle.

Of course average is not a perfect tool. If instead the student scores are

$$50, 50, 50, 50, 50, 50, 50, 50, 50, 50, 50, 50, 50,$$

$$100, 100, 100, 100, 100, 100, 100, 100, 100, 100, 100, 100$$

then the average is $\mu = 74$. A very similar average to the first example. But this value of μ does not tell us that 13 of the students did very poorly on the test while 12 of the students scored perfectly.

The concept of *standard deviation* has been invented to measure how the aggregate of data differs from the average value μ. If the standard deviation is small, then the data values are clustered around μ. If the standard deviation is large, then the data values are all over the place. We shall not take the time in this exposition to give a detailed treatment of standard deviation.

The subject of statistics is less than 100 years old. It is a relatively new science. But it is an increasingly important one. The spread of disease, the performance of the stock market, the configurations of a molecule, the growth of a tumor, the sales of a new product, are all governed and described by statistics.

Traditionally, statisticians dealt with small sets of data—typically not more than 100 numbers. In the early days this was because all the calculations had to be done by hand. Since we now have computers, it makes sense that we can handle larger data sets. But the motivation, in the early days, to handle large data sets was not great. This has all changed.

One of the many reasons for this change, but perhaps the first, is genetics. A human cell contains 46 chromosomes. Each chromosome has many thousands of genes and billions of base pairs.[1]

DNA profiling (also called DNA fingerprinting, DNA testing, or DNA typing) is a forensic technique used to identify individuals by characteristics of their DNA. A DNA profile is a small set of DNA variations that is very likely to be different in all unrelated individuals, thereby being as unique to individuals as are fingerprints (hence the alternative name for the technique). DNA profiling should not be confused with full genome sequencing. First developed and used in 1985, DNA profiling is used in, for example, parentage testing and criminal investigation, to identify a person or to place a person at a crime scene; these techniques are now employed globally in forensic science to facilitate police detective work and help clarify paternity and immigration disputes. Although 99.9% of human DNA sequences are the same in every

---

[1]Here a *base pair* is any of the pairs of nucleotides connecting the complementary strands of a molecule of DNA or RNA and consisting of a purine linked to a pyrimidine by hydrogen bonds. The base pairs are adenine-thymine and guanine-cytosine in DNA and adenine-uracil and guanine-cytosine in RNA or in hybrid DNA-RNA pairing.

person, enough of the DNA is different that it is possible to distinguish one individual from another, unless they are monozygotic ("identical") twins. DNA profiling uses repetitive ("repeat") sequences that are highly variable, called variable number tandem repeats (VNTRs), in particular short tandem repeats (STRs). VNTR loci are very similar between closely related humans, but are so variable that unrelated individuals are extremely unlikely to have the same VNTRs.

It should be stressed that DNA pairing is *not* like pairing socks. We are dealing here with billions of pieces of data, and statistical techniques must be used to organize the data. This is why it takes up to a week to do a DNA sequence and to compare the DNA of a criminal with DNA found at the scene of a crime. [Compare fingerprinting. Using modern computer and wavelet techniques, a suspect's fingerprint can be compared to fingerprints in a database in just a few seconds.]

The handling of DNA and the data that it generates, as we have been describing it, is an example of a new field of study called *big data*. In the past twenty years, big data has become a part of life, and it has entered the study of many different aspects of our existence. We describe two of them here.

The first concerns city planning. Issues that are present in any city, and that are amenable to analysis by big data techniques, include predictive policing, improving traffic flow, and promoting energy efficiency. Another concept that can be better understood with big data methods is the perception of how cities and neighborhoods grow and develop.

Where does one obtain the needed data? One obvious source is the Internet. `Google`, `Twitter`, and `Yelp` have provided data that give new insights into cities and urban change. `Google Street View` has been used by sociologists Robert Sampson and Jackelyn Hwang to examine the role of race in the process of gentrification of neighborhoods and in neighborhood transformation.

In addition, the U.K. Spatial Economics Research Centre has used geo-tagged photos from `Flickr` to estimate levels of urbanity in London and Berlin. Data on mobility from `Uber` and `Lyft` (and even from traditional taxicabs) has been used in some recent studies. Real estate sites such as `Zillow` and `Trulia` have provided data to analyze housing price trends across neighborhoods, cities, and metropolitan areas.

Reviewer data from `Yelp` has been used to study gentrification and unequal urban consumption patterns. Another study used `Yelp` review to understand the connection between gentrification and race in Brooklyn. `Yelp` found yet another use in finding out how ethnic and racial segregation affects consumption levels in New York City.

The Oxford Internet Institute used `Twitter` data to chart the flow of online content and ideas across cultures. The cartography blog `Floating Sheep` has implemented data from `Twitter`, `Google`, and `Wikipedia` to map consumption of beer and pizza as well as weed, bowling, and strip clubs. *My Space* has been used to track the leading centers for popular music genres across the United States and around the world.

A team of Italian researchers has combined data from `Foursquare` and `OpenStreetMap` to test Jane Jacobs's theories of urban vitality and diversity in 6 Italian cities. This study confirmed many of Jacobs's insights about the importance of short blocks, mixed land use, walkability, dense concentration of talented workers, and urban public space.

The study of big data requires a number of new ideas. Gunnar Carlson of Stanford University has devised topological methods (from a branch of pure mathematics) to manipulate large data sets. He has even started a company to develop these ideas. Machine learning techniques (as developed by mathematician and Fields Medalist Stephen Smale) are used to derive meaning from large sets of data. As an instance, the Livelihoods Project from the School of Computer Science at Carnegie Mellon University uses machine learning to analyze 18 million check-ins on `Foursquare` to determine the structure and characteristics of 8 different cities. Big data and new data analytics can help researchers determine urban structures and patterns that traditional data and methods might not discern.

A recent NBER study by Harvard and MIT researchers (see for instance [FHS]) uses computer visioning to better comprehend geographic differences in income and housing prices. They use `Google Street View` to predict income levels and housing prices in Boston and New York between 2007 and 2014. The study links 12,200 images of New York City and 3,600 images of Boston to data on median family income and home values from the years 2006–2011 from the American Community Survey. It determines the extent to which the positive physical attributes shown in these images (things like size and green space) attract more affluent residents and predict incomes and housing prices. In the end this study has determined that "images can predict income at the block group level far better than race or education does." The study asserts that a key purpose of big data is to help illuminate the role of smaller geographic area in the urban economy; these are more difficult to get at with traditional census data. In sum, big data offers "some hope that `Google Street View` and similar predictors will enable us to better understand patterns of wealth and poverty worldwide."

Our second example of big data comes from the study of a global early warning system for infectious diseases. The idea is to use computer models to tap into environmental, epidemiological, and molecular data, gathering the intelligence needed to forecast where disease risk is high and what actions could prevent future outbreaks or contain epidemics.

The goal here is to shift the infectious disease paradigm from reactive—where first responders struggle to contain active threats, as in recent Ebola and Zika outbreaks—to pre-emptive management of risk. Intelligence on infectious disease could assess vulnerabilities based on the flow of risk in real time, and thereby inform targeted responses that will minimize damage.

We mathematicians have been fond of telling people, for many years now, that they need to know some mathematics in order to cope with modern life. In particular, it is essential for survival to be able to understand the spread of disease and the threat that the disease poses. Unfortunately, most people pay no attention to these admonitions, and those few that do rarely attain the needed level of expertise. The goal of the projects being described here is to sidestep the responsibility of the individual to protect his/her health and to find ways to anticipate and therefore eliminate the spread of disease.

As expert Barbara Han says, "For far too long our main strategy for tackling infectious disease has been defense after emergence, when a lot of people are already suffering. We are at an exciting point in time where technology and Big Data present us with another option, one that is anticipatory and has real potential to improve global health security." See [HPJA] for further details.

Han and her collaborator John Drake propose a 3-tiered system with "watches," "warnings,", and "emergencies." Like the commonly used system for weather alerts, this would help decision makers and the public to make more informed decisions. As they assert [HAN], "Much of the destructive potential of infectious diseases stems from the fact that they often strike unexpectedly, leaving little time for preparation. The best countermeasure is therefore an early warning to give affected regions or communities more time to prepare for impact."

Machine learning methods have been used to mine data from multiple sources to identify animal species that are likely to carry disease and also geographic hotspots vulnerable to outbreaks of specific pathogens. Ramping up this effort to create a tool for global health authorities will require an increase in the stream of data available for modeling, investment in a quantitative workforce, and open dialogue among academic modelers and decision makers.

Drake [HAD] has observed that, "One of the key problems is figuring out how to integrate multiple data streams. Also, there are some aspects of epidemics that are poorly understood and can change quickly, such as how individual behavior changes in the face of a perceived acute health threat. We need better sources of information about these processes, if we are to develop a reliable basis for forecasting."

A global early warning system for infectious disease would transform efforts to advance global health security and would improve global health equity. It is an exciting new development, and one from which we can all benefit.

Exercise: One of the important sources for big data problems is genetic. One such example is gene matching. Read up on this topic and explain how and why big data plays a role.

Chapter 6

Counting

6.1 Funny Numbers

Let us determine how many zeros end the number 100!

Recall that
$$100! = 100 \cdot 99 \cdot 98 \cdots 3 \cdot 2 \cdot 1.$$

Adding a zero to the end of a product occurs precisely when we multiply by 10. Thus multiplication by any number ending in 1, 3, 7, 9 cae nnot possibly add a zero to the product (since none of these numbers divides 10). In fact the prime factorization of 10 is $10 = 5 \cdot 2$. We endeavor to solve this problem by counting the factors of 5 in 100!

In the numbers 1–10, only the numbers 5 and 10 have factors of 5. The number 5 may be paired with 2 to yield 10 and the number 10 does not need to be paired. The two resulting factors of 10 contribute two zeros to the full product that forms the factorial.

In the numbers 11–20, only the numbers 15 and 20 have factors of 5. Reasoning as in the last paragraph, we count two additional zeros.

The numbers between 21 and 30 are a bit different. As before, 25 and 30 are the only numbers having a factor of 5, but 25 has *two* factors of 5. Thus
$$22 \times 24 \times 25 = 11 \times 12 \times (2 \times 5) \times (2 \times 5)$$
and this will contribute $10 \times 10 = 100$, or two zeros. Thus, together with the

one zero from the number 30, the range 21–30 contributes a total of three zeros.

The range 31–40 is a simple one, like the first two ranges we considered. It contributes two zeros.

The range 41–50 is special because 45 contributes one factor of 5 but 50 contributes two factors. Thus this range of numbers contributes three zeros (as did the range 21–30).

The range 51–60 and the range 61–70 are like the first two. There are no multiple factors of 5, and each range contributes two zeros.

The range 71–80 is special because 75 contributes two factors of 5 and 80 contributes one factor of 5. The total contribution is three zeros.

The range 81–90 contributes two factors of 5 in the usual fashion, and thus adds two zeros.

The range 91–100 contains 95 and 100. The first of these contributes one factor of five and the second contributes two. Thus three zeros are added.

Taking all of our analyses into account, we have six ranges of numbers that each contribute two zeros and four ranges that each contribute three zeros. This gives a total of 24 zeros that will appear at the end of 100!

Exercise: What is the last digit of 143^{100}? What are the last two digits of 277^{1000}?

6.2   The Pigeon Flew the Coop

In this section we discuss a very simple and elegant mathematical idea. Indeed it is so simple that you could explain it to a little sister or to your grandmother. And they would understand immediately and without question. Nevertheless, this is a very powerful and useful idea that has seen dramatic applications in the mathematical sciences.

Discovered originally in 1834 by Peter Gustav Lejeune Dirichlet (1805–1859) (and originally known as the *Dirichletscher Schubfachschluss* or "Dirichlet's drawer-shutting principle"), we now call this idea the *pigeonhole principle*. And it says this.

Figure 6.1: The pigeonhole principle.

Suppose that you deliver $(n + 1)$ letters to n mailboxes. Then some mailbox will contain at least two letters. See Figure 6.1.

Pretty simple, no? And the reason it is true is also simplicity itself. Suppose that the assertion is false. Then each of n mailboxes contains at most 1 letter. But that means that the total number of letters is at most n. This contradicts the fact that there are actually $(n + 1)$ letters. So the pigeonhole principle is established.

What is astonishing about this little idea is that it has such diverse and notable applications. We describe some of them now.

A Hairy Problem

It is a fact that there are 2 people in London with the same number
of hairs on their heads. How can we know this?

The population of London is 8.5 million. No person has more
than 1 million hairs on his head. Think of each person as a letter,
and that person is going to be delivered to a mailbox with a number
on it corresponding to the number of hairs on the head. So we are
delivering 8.5 million letters to 1 million mailboxes. We apply the
pigeonhole principle with $n = 1$ million. The conclusion then is that
some mailbox contains at least two letters. That means that there are
at least 2 people with the same number of hairs on their head.

Note that, in this last example, we are not simply saying that there are
two Londoners with 587,294 hairs on their head and that is the end of it.
In fact it is fairly simple to see that there is some number of hairs that 8
Londoners have.

Party Politics

A party is attended by 50 people. A great variety of handshaking takes place. Let us demonstrate that there are two people who shake the same number of hands.

Let us imagine that there are mailboxes, each labeled with a number from 0 to 49. The name of party attendee A is put into mailbox n if A shakes n hands.

Now it is important to note that either box 0 or box 49 is empty. Because if someone shakes 49 hands then there is nobody who shook 0 hands. And if somebody shook 0 hands then there is nobody who shook 49 hands. So there are really only 49 mailboxes into which we can put the 50 names. It follows from the pigeonhole principle that some mailbox contains 2 names. Therefore there are 2 people who shook the same number of hands.

Definitely a Set-To

Consider the set $S = \{1, 2, 3, 4, 5, 6, 7, 8, 9\}$. We claim that any subset of S with 6 elements must have two elements in it that add up to 10.

To see this, let us first set up the mailboxes. There will be five mailboxes with these labels:

$$\{1, 9\} \ , \ \{2, 8\} \ , \ \{3, 7\} \ , \ \{4, 6\} \ , \ \{5\} \, .$$

Now let X be a subset of S with 6 elements. We put each element into a mailbox that has the identity of that element in its label. For example, if X contains 4, then 4 will go into the fourth mailbox because that mailbox has 4 in its label.

If X has 5 in it, then 5 goes into the fifth and last mailbox because that box has 5 in its label.

Well, there are 6 elements of X going into 5 mailboxes, so some mailbox will contain two elements. Those elements will be different, and they will be the same as the two labels of the mailbox. So they will add to 10. [Note here that the last mailbox can contain only one element—namely 5 itself—so it is irrelevant to this discussion.]

Sock It to Me

Suppose that you have some red socks, some blue socks, some green socks, some black socks, and some brown socks. How many socks must you have in order to be certain that you have a pair?

There are 5 different colors of socks. So 6 socks will do the trick. Think of it this way. You have 5 mailboxes, labeled red, blue, green, black, and brown. You stick each sock that you own into a box whose label is the color of the sock. Since there are 6 socks and 5 mailboxes, some box will end up containing two socks. And they will be of the same color. So they form a pair.

Let us observe that two sets S and T have the same number of elements if and only if there is a one-to-one correspondence between the elements of the sets. For instance,

$$S = \{1, 2, 3, 4, 5, 6\}$$

and

$$T = \{a, b, c, d, e, f\}$$

each have the same number of elements, namely 6. And there is a one-to-one correspondence between the elements of the two sets:

$$1 \leftrightarrow a$$
$$2 \leftrightarrow b$$
$$3 \leftrightarrow c$$
$$4 \leftrightarrow d$$
$$5 \leftrightarrow e$$
$$6 \leftrightarrow f$$

On the other hand, the pigeonhole principle tells us that if the two sets have *different* numbers of elements then there is no one-to-one correspondence between them. Just because one set, say S will have more elements than T. So any correspondence of S to T will send two elements of S to the same place in T.

We can illustrate this last set of ideas with a nice example.

Checkmate

Imagine a chessboard as shown in Figure 6.2. We remove two of the diagonally opposite squares from the board, as shown in Figure 6.3. Now we have a collection of dominoes as shown in Figure 6.4. Each domino will exactly cover two squares of the chessboard. The question is: Can we cover the entire deleted chessboard, as shown in Figure 6.3, with dominoes?

Let us think about this matter analytically. The original chessboard had 8×8 or 64 squares. So the deleted chessboard has 62 squares. It would require 31 dominoes to cover all the squares of this deleted board. What is important to observe is that each domino will cover one white square and one gray square. So covering the deleted board by dominos will create a one-to-one correspondence between white squares and gray squares on the deleted board.

The trouble is that the two squares that we removed are both gray (compare Figures 6.2 and 6.3). So the deleted board has 32 white squares and $32 - 2 = 30$ gray squares. According to our pigeonhole principle discussion preceding this example, there cannot be a one-to-one correspondence between the white squares and the gray squares on the deleted board. Thus it is impossible to cover the deleted board with dominoes.

Exercise: Explain why this more general version of the pigeonhole principle is true: For natural numbers k and m, if $n = km + 1$ objects are distributed among m sets, then at least one of the sets will contain at least $k + 1$ objects.

6.3   Conditional Probability

One of the big ideas in probability theory is that of conditional probability. This is best illustrated by an example.

Consider the problem of flipping two coins. What is the probability that they will both come up heads? Well, there are four possible outcomes as illus-

Figure 6.2: A standard chessboard.

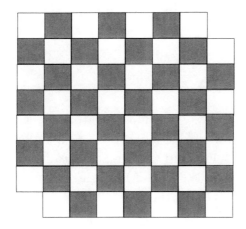

Figure 6.3: The deleted chessboard.

Figure 6.4: A domino.

trated in the following display (with H standing for "heads" and T standing for "tails"):

$$TT \,,\; TH \,,\; HT \,,\; HH \,.$$

We see that in only one of four instances do we come up with two heads. So the chances are 1 in 4, or .25.

But what if we instead say, "Assume that one coin is already flipped and is a head. What is the chance that the second one will be a head also?" That is a completely different question. The second coin can come up heads or tails and those are the only two possibilities. This is independent of what the first coin did. So the odds are 1 in 2 or .5 that the second coin will come up heads.

This last is an example of conditional probability. It is probability calculated on the basis of a condition. A famous formula of Bayes enables us to calculate conditional probability in practice. It says that

$$P(A|B) = \frac{P(B|A) \cdot P(A)}{P(B)} . \qquad (6.3.1)$$

Here $P(A|B)$ is the probability of A given B and $P(B|A)$ is the probability of B given A.

Example: Let us apply formula (6.3.1) to the situation of two coins, where one coin is already known to be a head. So A is "What is the probability of two heads" and B is "One coin is already a head." Thus, as we have seen, $P(A|B)$ is .5. Also (trivially) $P(B|A)$ is 1. Finally, $P(A)$ is .25 and $P(B)$ is .5. So Bayes's formula tells us that

$$.5 = \frac{1 \cdot .25}{.5}$$

and that is certainly true.

Our first example may make this whole matter seem rather silly, but it is not. There are some truly fascinating applications that we shall now discuss.

As a first instance, let us discuss the likelihood that a patient will have cancer. There is some general sense that older people are more likely to get cancer than younger people. Let us explore that assertion. Empirical observation suggests that 1% of the population will have cancer. So let us accept that as a hypothesis.

Now let us think about a patient who is 65 years old. We would like to know the likelihood of a 65-year-old person having cancer.

It is an observed fact that the probability that a randomly chosen person is 65 years old is 2%. Also the probability that a person *with cancer* is 65 years old is 5%. Note that this number is certainly greater than the previously cited 1%.

Now let us use Bayes's Law (6.3.1) to calculate the probability that a person who is 65 years old will have cancer. We let B be the assertion that a person is 65 years old and A be the assertion that a person has cancer. Then we see that

$$P(A) \ = \ 1\%$$
$$P(B) \ = \ 2\%$$
$$P(B|A) \ = \ 5\%$$

Thus we may calculate that

$$P(A|B) = \frac{P(B|A) \cdot P(A)}{P(B)} = \frac{.05 \cdot .01}{.02} = \frac{.0005}{.02} = .025 \, .$$

Thus we see that the probability that a 65-year-old person will have cancer is 2.5%. That is not at all obvious from the given data. In fact the result is lower than we might have anticipated. The reason is that the base rate for cancer is 1%, and that is itself rather low.

The next example is even more surprising. As you know, we are frequently told when we are given a medical test that, "This test is 90% accurate." Thus, thinking intuitively, we are led to believe that whatever report the physician ultimately gives us has a 90% chance of being correct. If the physician tells us that the report came out positive, then we think that we have a 90% chance of having the disease. If instead the physician tells us that the report is negative, then we think that we have a 90% chance of not having the disease. Bayes's theorem will show us that in fact this supposition is completely incorrect.

Suppose that we have a test for drug users. It is known to be 99% sensitive and 99% specific. This means, in plain English, that when the test is applied to a drug user, it will come out true positive 99% of the time. And when it is a applied to a non-user, it will come out true negative 99% of the time.

Let us use the notation $P(+|\text{User})$ to denote the probability that a person tests positive given that he/she is a drug user. Likewise we use the notation $P(+|\text{Non-user})$ to denote the probability that a person tests positive given that he/she is a non-user. We shall assume that 0.5% of people are actually drug users. Then Bayes's theorem tells us that

$$P(\text{User}|+) = \frac{P(+|\text{User}) \cdot P(\text{User})}{P(+)}.$$

For the sake of this calculation, we find it convenient to write

$$P(+) = P(+|\text{User}) \cdot P(\text{User}) + P(+|\text{Non-user}) \cdot P(\text{Non-user}).$$

Thus we see that

$$P(\text{User}|+) = \frac{P(+|\text{User}) \cdot P(\text{User})}{P(+|\text{User}) \cdot P(\text{User}) + P(+|\text{Non-user}) \cdot P(\text{Non-user})}.$$

Now we know each of the probabilities on the right-hand side. So we may calculate that

$$P(\text{User}|+) = \frac{0.99 \cdot 0.005}{0.99 \cdot 0.005 + 0.01 \cdot 0.995} \approx 0.332.$$

Astonishingly, we find that the likelihood that a person is a drug user, given that he/she has tested positive, is just 33.2%. In other words, if the person tests positive, it is more likely that he/she is *not* a drug user. Again, this surprising result is determined by the base rates. Put more explicitly, the number of non-users is very large compared to the number of users; hence the number of false positives (0.995%) outweighs the number of true positives (0.495%). Let us illustrate the idea with specific numbers. Suppose that 1000 people are tested. We thus expect there to be 995 non-users and 5 users. From the 995 non-users, we expect $0.01 \times 995 \approx 10$ false positives. From the 5 users, we expect $0.99 \cdot 5 \approx 5$ true positives. So we see that, out of 15 positive tests, only 5, or about 33% are genuine.

As an exercise, you may wish to calculate what happens if the sensitivity is 99% and the specificity is 99.5%. Then the probability that a person with a positive test is a drug user rises to 49.8%.

Now let us completely shift gears and look at the output from a factory. This particular factory has just three machines. Call them A_1, A_2, and A_3. These machines account for 20%, 30%, and 50% of the output respectively. The percentage of defective products produced by the machines is

$$A_1 \text{ produces } 5\% \text{ defective}$$
$$A_2 \text{ produces } 3\% \text{ defective}$$
$$A_3 \text{ produces } 1\% \text{ defective}.$$

If an item is chosen at random from the total output one day, and that item is found to be defective, then what is the likelihood that that item was produced by machine A_3?

By way of a solution to this problem, let us let α_j denote the event that a randomly chosen item was made by machine j (for $j = 1, 2, 3$). Let β denote the event that an item is defective. Obviously we are interested in $P(\beta|\alpha_j)$ and $P(\alpha_j|\beta)$.

Now we are given that

$$P(\alpha_1) = 0.2 \ , \quad P(\alpha_2) = 0.3 \ , \quad P(\alpha_3) = 0.5 \ .$$

We also know that

$$P(\beta|\alpha_1) = 0.05 \ , \quad P(\beta|\alpha_2) = 0.03 \ , \quad P(\beta|\alpha_3) = 0.01 \ .$$

In order to apply Bayes's theorem, we need to calculate $P(\beta)$. We see that

$$
\begin{aligned}
P(\beta) &= P(\beta|\alpha_1) \cdot P(\alpha_1) + P(\beta|\alpha_2) \cdot P(\alpha_2) + P(\beta|\alpha_3) \cdot P(\alpha_3) \\
&= (0.05)(0.2) + (0.03)(0.3) + (0.01)(0.5) = 0.024 \ .
\end{aligned}
$$

So we have found that 2.4% of the total factory output is defective.

Our job now is to calculate $P(\alpha_3|\beta)$. Using Bayes's Law, we find that

$$P(\alpha_3|\beta) = \frac{P(\beta|\alpha_3) \cdot P(\alpha_3)}{P(\beta)} = \frac{(0.01)(0.50)}{(0.024)} = 0.2083 \ .$$

In conclusion, given that an item is defective, the probability that it was produced by machine A_3 is only about 5/24. This is surprising, for machine

A_3 produces half the output, but it produces only about a fifth of the defective items.

Again we may endeavor to understand this situation by examining an instance with concrete numbers. Suppose that 100,000 items are produced by the factory. So 20,000 will be produced by machine A_1, 30,000 will be produced by machine A_2, and 50,000 will be produced by machine A_3. Now notice that machine A_1 will produce 1,000 defective items, machine A_2 will produce 900, and machine A_3 will produce 500. So we see that, of the total $1000 + 900 + 500 = 2400$ defective items, only 500 are produced by the third machine. That is 5/24 of the total number of defective items.

Exercise: Suppose that somebody secretly rolls two fair six-sided dice, and we must predict the outcome (the sum of the two upward faces). Let A be the value rolled on die 1. Let B be the value rolled on die 2. What is the probability that $A = 2$ given that $A + B = 5$?

6.4 Benford's Law

The phenomenon that we shall describe here was first observed by Simon Newcomb in 1851. The physicist Frank Benford noted the same effect in 1938, and performed some experiments to confirm it. So our discussion here also illustrates another much-studied phenomenon: that new ideas are often named after the second person to discover them, not the first.

Benford's Law concerns the nature of whole numbers that we encounter in everyday life. Suppose you are looking at the populations of the states in the United States, or the heights of the 60 tallest buildings in the world, or the (numerical) street addresses of the first 500 people listed in *American Men and Women of Science*, or the number of bacteria in a petri dish plotted over a period of 50 days. Each of these is a list of integers. Now there are ten distinct digits in our number system: 0, 1, 2, 3, 4, 5, 6, 7, 8, 9. Which of these will most frequently occur as a *first* digit?

Well, 0 never occurs as a first digit. So we can forget about 0. So there are 9 other choices. And there is no particularly obvious reason why any one of them should occur with any greater frequency than any other of them. So we might guess that each of the digits 1, 2, 3, 4, 5, 6, 7, 8, 9 occurs with equal likelihood as a first digit. That is to say, each of these nine digits occurs 11.1% of the time.

Surprisingly, this is not the case. Benford's Law says that 1 occurs as a first digit about 30% of the time. And 9 occurs as a first digit only about 5% of the time. How could this be?

We shall do two things in this discussion: **(a)** we shall offer a scientific reason for Benford's Law and **(b)** we shall give several objective examples to illustrate Benford's Law.

First, why should Benford's Law be true? As you know, thanks to the Arabs, we do our arithmetic base 10. That is to say, the digits in any positive integer represent powers of 10. Just as an instance, the number

$$3648$$

represents 8 units of 10^0, 4 units of 10^1, 6 units of 10^2, and 3 units of 10^3. If we want a 4-digit number n to begin with the digit 1, then we want n to be a power of 10 that is at least 3 (because 10^3 is 1000) and not more than 3.301 (because $10^{3.301}$ is 1999). The length of this range is $\ell_1 = 3.301 - 3 = 0.301$.

Now contrast this last calculation with the situation if we want a 4-digit number n to begin with the digit 9. Then we want n to be a power of 10 that is at least 3.954 (because $10^{3.954}$ is 9000) and not more than 3.999 (because $10^{3.999}$ is 9999). The length of this range is $\ell_2 = 3.999 - 3.954 = 0.045$.

Now we see clearly that ℓ_1 is much greater than ℓ_2. In fact

$$\frac{\ell_1}{\ell_2} = 6.69 \, .$$

In other words, it is more than 6 times more likely that a four-digit integer will begin with 1 than it is that the four-digit integer will begin with 9. The key is to notice that the actual numbers in use by humans tend to follow frequencies related to orders of magnitude.

You may recall from above that we said that a first digit of 1 occurs with likelihood about 30% while a first digit of 9 occurs with a likelihood of less than 5%. That contention is consistent with the calculations that we just did.

Enough said about the scientific reason for Benford's Law. If you are interested, you can do the calculations to see how likely it is that a 4-digit integer will begin with 5. Or 8. Now we shall move on to looking at some interesting examples.

Let us begin by examining the heights of the world's tallest buildings. As you can see in the table below, we have specified the heights of the 22 tallest buildings. We have stopped once we reached height 787.4, just because none of the heights after that will begin with 1. Of the 22 heights we have specified, 2 of them begin with 2 and 18 of them begin with 1. One begins with 7 and one begins with 9. So Benford's Rule holds decisively in this situation. Of course one could argue that it is obvious that most of the tallest buildings will be in the 1000-foot range. But we will look at other, possibly more interesting, examples below.

Building	Country	Height Ft.
Burj Khalifa	United Arab Emirates	2722
Tokyo Skytree	Japan	2080
Abraj Bait Towers	Saudi Arabia	1972
INS Kattabomman	India	1545
VLF Transmitter	United States	1503
Petronas Twin Towers	Malaysia	1482
Empire State Building	United States	1454
Sears Tower	United States	1451
432 Park Avenue	United States	1396
Trump Hotel and Tower	United States	1386
GRES-2 Power Station	Kazakhstan	1377
Dimona Radar Facility	Israel	1312
Kiev TV Tower	Ukraine	1263
Zhoushan Tower Overhead Powerline	China	1214
Gerbrandy Tower	Netherlands	1203
TV Tower Vinnytsia	Ukraine	1161
Millau Viaduct	France	1122
Sky Tower	New Zealand	1076
Amazon Tall Tower Observatory	Brazil	1066
Lakihegy Tower	Hungary	1031
Jinping-I Dam	China	1001
Samatra TV Tower	India	984
Moscow State University	Russia	787.4

Next let us consider the populations of the states in the United States.

State	2015 Population
California	39, 144, 818
Texas	27, 469, 114
Florida	20, 271, 272
New York	19, 795, 791
Illinois	12, 859, 995
Pennsylvania	12, 802, 503
Ohio	11, 613, 423
Georgia	10, 214, 860
North Carolina	10, 042, 802
Michigan	9, 022, 506
New Jersey	8, 958, 013
Virginia	8, 382, 993
Washington	7, 170, 351
Arizona	6, 828, 065
Massachusetts	6, 794, 442
Indiana	6, 619, 680
Tennessee	6, 600, 299
Missouri	6, 083, 672
Maryland	6, 006, 401
Wisconsin	5, 771, 337
Minnesota	5, 489, 594
Colorado	5, 456, 574
South Carolina	4, 896, 146
Alabama	4, 858, 979
Louisiana	4, 670, 724
Kentucky	4, 425, 092
Oregon	4, 028, 977
Oklahoma	3, 911, 338
Connecticut	3, 590, 886
Iowa	3, 123, 899
Utah	2, 995, 919
Mississippi	2, 992, 333
Arkansas	2, 978, 204
Kansas	2, 911, 641
Nevada	2, 890, 845
New Mexico	2, 085, 109

State	2015 Population
Nebraska	1,826,341
West Virginia	1,853,011
Idaho	1,654,930
Hawaii	1,431,603
New Hampshire	1,330,608
Maine	1,329,328
Rhode Island	1,056,298
Montana	1,032,949
Delaware	945,934
South Dakota	858,469
North Dakota	756,927
Alaska	738,432
Vermont	626,042
Wyoming	586,107

It is easy to see that there are 50 states, and 8 of them with populations having first digit 1. That is *not* 30%; instead it is 16%. But, by contrast, the number of states having population with first digit 9 is 1. The number of states having population with first digit 8 is 2. The number of states having population with first digit 7 is 1. So the digit 1 is still the winner. We conclude by noting that the number of states having population with first digit 2 is 8 and the number of states having population with first digit 3 is 4 and the number of states having population with first digit 4 is 5 and the number of states having a population with first digit 6 is 6. So the number of hits as we go from 1 to 2 on down to 9 is steadily, although not monotonically, decreasing.

The example of the populations of states is a bit more interesting than the example of heights of buildings because there is no a priori upper bound on the populations of states. The field is more open.

Benford's Law has been used in real life to detect fraud. Because people who create fraudulent data tend to distribute the digits randomly; they are unaware of Benford's Law. So it is thereby easy to pick out sets of fraudulent data. As a result, Benford's Law has been used in forensic accounting and auditing as an indicator of accounting and expense fraud. In the United States, evidence based on Benford's Law has been admitted to court in criminal cases at the federal, state, and local levels.

Benford's Law was invoked as evidence of fraud in the 2009 Iranian national elections. It has also been used to analyze other election results.

Statisticians routinely use Benford's Law to test the legitimacy of various data sets.

Let us examine the Fibonacci sequence and see whether Benford's Law applies to it. As you may know, the first two terms of the Fibonacci sequence are 1 and 1. Subsequent terms are derived by adding the preceding two terms. So the sequence is

$$1 , 1 , 2 , 3 , 5 , 8 , 13 , 21 , 34 , 55 , 89 , 144 , 233 ,$$

$$377 , 610 , 987 , 1597 , 2584 , \ldots.$$

You can see that we have written 18 terms of the Fibonacci sequence, and 5 of them have initial digit 1. We see that 4 of them have initial digit 2, 3 of them have initial digit 3, none of them has initial digit 4, 2 of them has initial digit 5, 1 of them has initial digit 6, none of them has initial digit 7, 2 of them has initial digit 8, and 1 of them has initial digit 9. While the Fibonacci sequence does not adhere strictly to the dictum of Benford's Law, it follows the spirit of Benford's Law.

A professor of probability likes to give the following exercise to his class on the first day. He says to the students, "This evening, if your mother's maiden name begins with a letter between *A* and *L*, flip a coin 200 times and record the result. If, instead, your mother's maiden name begins with a letter between *M* and *Z*, then fake a sequence of 200 heads and tales and record it on a piece of paper. Hand all these papers in to me tomorrow."

After collecting the papers, the professor is easily able, using the philosophy of Benford's Law, to recognize papers from the first group and papers from the second group. How does he do it? He is a probabilist, and he knows that, in a truly random sequence of 200 coin tosses, the probability of getting 6 heads or 6 tails in a row is *very high* (the mathematical proof of this assertion is rather complicated, and we omit it). But a person faking a sequence of 200 coin tosses is quite unlikely to put in a sequence of 6 heads or 6 tails in a row. End of story.

Accountants use this point of view to detect fraud. A typical collection of numbers on a page of a tax return will have a preponderance of numbers beginning with 1, somewhat fewer beginning with 2, somewhat fewer still that begin with 3, and so forth—with the rarest initial digit being 9. But people faking data tend not to know this, and they manufacture numbers with an evenly distributed selection of first digits. A great many cases of fraud have been detected in this fashion. In fact accountant Mark Nigrini has worked with several companies in New York, and is now consulting with the tax agencies of several countries, to develop fraud-detection techniques using Benford's Law.

Benford's Law is used in computer design. Knowing which digits occur most frequently allows a computer scientist to tailor storage algorithms so as to get the most data in the least space.

Exercise: Two friends play a game. They each secretly write a 5-digit integer on a piece of paper. Then they compare numbers. If the second digits match, then player A wins. If the second digits do not match, then player B wins. What is the best strategy for player A? What is the best strategy for player B?

6.5 Puzzling Birthdays

Here is an interesting combinatorial question. Find the least positive integer N so that, in a room containing N people, the odds are better than even that two of the people have the same birthday (not the year, just the day). [**Hint:** Forget about leap year. Assume that a year has 365 days.]

As has been indicated previously, it is sometimes convenient to calculate the probability that something will *not* happen, and then subtract that result from 1.

With this in mind, we fix N and determine the probability that no two people, in a room of N people, have the same birthday. Say that the people are $P_1, P_2, \ldots, P_N$. Now person P_1 may have any of the 365 days of the year for his/her birthday without contradicting the condition that no two people in the room will have the same birthday. Once person P_1 has fixed a birthday (this is usually fixed on the day that he/she is born, but we are rewriting history for the sake of our analysis), then person P_2 is not allowed to have that birthday if we are to maintain distinct birthdays. Thus person P_2 has 364 choices for his/her birthday.

And so it goes. If we are to maintain distinct birthdays, then person P_3 has 363 choices for his/her birthday. In summary, the total number of birthday combinations for N people, with no two being the same, is

$$365 \cdot 364 \cdot 363 \cdots [365 - (N - 1)].$$

The total number of *all possible* birthday distributions among N people, without regard for duplication or lack of duplication, is

$$\underbrace{365 \cdot 365 \cdot 365 \cdots \cdots 365}_{N \text{ times}}.$$

In summary, the probability that N people in a room will all have distinct birthdays is

$$P = \frac{365 \cdot 364 \cdot 363 \cdots [365 - (N - 1)]}{365^N}.$$

For the purposes of calculation, so that we do not have to deal with huge numbers, we rewrite this formula as

$$P = \frac{365}{365} \cdot \frac{364}{365} \cdot \frac{363}{365} \cdots \frac{365 - (N - 1)}{365}.$$

And now we must bring out our computer or calculator and begin to calculate. Begin at the left, multiplying the fractions together. When your product falls below $1/2$ you are finished. The last fraction that you multiplied in will tell you what N should be (because the last fraction that you multiplied in was $\frac{365 - (N-1)}{365}$).

The author has performed this calculation. He multiplied together 23 terms to obtain a probability of .4927027. Using only 22 terms yielded a probability of .5243046. Clearly the least N to yield a probability of less than half that each person in the room will have a distinct birthday is $N = 23$.

We conclude that if there are 23 people in a room, then the odds are better than even (in fact the odds are $P = 1 - 0.4927027 = .5072973$) that two of them will have the same birthday.

Exercise: How many people do you need in a room so that the probability is greater than 1/2 that three people will have their birthday on the same day?

Chapter 7

Games

7.1 How to Count

We begin our chapter on games with a prefatory section on the art of counting. This material has its own intrinsic interest, but will also be of use to us at several junctures.

Of course everyone knows how to count in some basic sense. If I hand you a bunch of grapes and ask how many there are, then you can count them and give me an answer. But there are many situations where counting is a challenge.

As an example, suppose you have seven wooden blocks that are seven different colors. You lay them out in a row—see Figure 7.1.

In how many different orders can you display the blocks? Figure 7.2 shows three possible orderings. But those are certainly not all the orderings. How many are there altogether?

We can analyze the question in this way. How many different blocks can be put in the first position (the one furthest to the left)? The answer is obviously 7. Any one of the seven different colored blocks can be put in

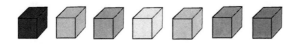

Figure 7.1: Seven blocks in a row.

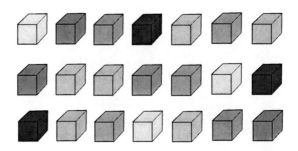

Figure 7.2: Three different orderings of the blocks.

that first position. Now how many different blocks can be put in the second position? Well, one block has been used in the first position. There are six blocks left. So there are 6 possibilities for the second position. Continuing in this fashion, we see that there are 5 possible blocks for the third position, 4 possible blocks for the fourth position, and so forth down to just 1 block for the seventh position. Thus the total number of possible arrangements of the blocks is

$$T = 7 \cdot 6 \cdot 5 \cdot 4 \cdot 3 \cdot 2 \cdot 1.$$

This is a number that is commonly denoted by 7! and called *7 factorial*. With a calculator we can quickly determine that it is equal to 5040.

More generally, suppose that we have k wooden blocks, where k is some positive integer. In how many different ways can we arrange these k blocks? In analogy with what we said in the last paragraph, we see that the number of possibilities is

$$k \cdot (k-1) \cdot (k-2) \cdots \cdots 3 \cdot 2 \cdot 1.$$

This number is commonly denoted by $k!$ and called k *factorial*.

Now here is a philosophically related, but more subtle, question. Suppose that we are given k wooden blocks. In how many different ways can we choose n blocks from among those k (here of course we assume that n is not greater than k). This question is best understood by way of an example.

Consider a collection of 3 blocks, as shown in Figure 7.3. In how many different ways can we choose 2 blocks from among those 3?

There are three possibilities:

(i) We could choose the first and second blocks.

Figure 7.3: Three wooden blocks.

(ii) We could choose the first and third blocks.

(iii) We could choose the second and third blocks.

And those are the only possibilities. There are three different ways to choose 2 blocks from among 3 blocks.

We would like now to understand the more general situation. Suppose that we are given k boxes and we want to choose n of them. A simple way to do so would be to follow these steps:

(i) Pick some ordering of the k boxes.

(ii) Now select the first n of the boxes in the new arrangement.

Let us illustrate the idea with choosing 2 boxes from among 3. Look at Figure 7.4. This time we have numbered the boxes instead of coloring them.

The figure shows all 6 possible arrangements of the boxes. Selecting the first two boxes from each arrangement, we see a certain redundancy:

$$
\begin{array}{c|c}
1 & 2 \\
1 & 3 \\
2 & 1 \\
2 & 3 \\
3 & 1 \\
3 & 2 \\
\end{array}
$$

The selection of $\{1,2\}$ in row 1 is repeated by the selection $\{2,1\}$ in row 3. The selection of $\{1,3\}$ in row 2 is repeated by the selection $\{3,1\}$ in row 5. And the selection of $\{2,3\}$ in row 4 is repeated by the selection of $\{3,2\}$ in row 6.

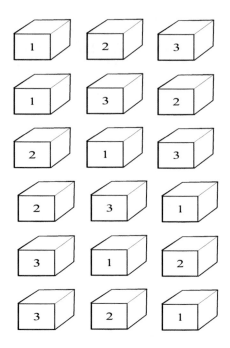

Figure 7.4: Six different orderings for 3 wooden blocks.

What we have learned is that we cannot say that there are 6! ways to choose 2 boxes from among three. We have to divide out by the fact that those first two boxes have two different orderings. So the correct answer is $3!/2! = (3 \cdot 2 \cdot 1)/(2 \cdot 1) = 3$.

In general, when we are choosing n boxes from among k boxes, we can look at all the different orderings of k boxes and choose the first n of these. But we must divide out by the fact that there are $n!$ ways to order those first n boxes. And there are also $(k - n)!$ ways to order the last $(k - n)$ boxes. In sum, the number of ways to choose n boxes from among k boxes is

$$\frac{k!}{n! \cdot (k - n)!}.$$

This quantity is so universally used in mathematics that we have a special notation for it. We write

$$\binom{k}{n} = \frac{k!}{n! \cdot (k - n)!}.$$

The expression $\binom{k}{n}$ is called the *choose function*.

In the next section we shall make good use of these ideas to analyze the lottery.

Exercise: Pascal's Triangle is the array

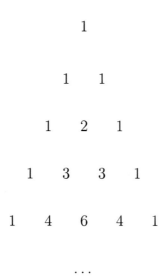

Notice that each non-end-position entry in row 3 and beyond is the sum of the two entries above it. Explain what the numbers in Pascal's triangle have to do with the choose function.

7.2 How to Beat the Lottery

Many of us buy lottery tickets with the hope of becoming instantly rich without any effort. It is rather foolish—because the odds are millions to one against us—but it is fun and we do it.

In the state of New York there is a lottery game called Quick Draw. Similar to the Las Vegas game of Keno, it is played like this. There is an array of the first 80 integers—see the displayed table.

1	2	3	4	5	6	7	8	9	10
11	12	13	14	15	16	17	18	19	20
21	22	23	24	25	26	27	28	29	30
31	32	33	34	35	36	37	38	39	40
41	42	43	44	45	46	47	48	49	50
51	52	53	54	55	56	57	58	59	60
61	62	63	64	65	66	67	68	69	70
71	72	73	74	75	76	77	78	79	80

You select m numbers from this array, where $1 \leq m \leq 10$. The state, which runs the lottery, picks 20 numbers from the array. And you are paid off according to how many of your m numbers match the state's 20 numbers.

A first natural question to ask is "What is the probability that I will get p matches?" Here of course $1 \leq p \leq m$.

We begin by calculating in how many different ways we can get p matches from among the 20 numbers that the state has chosen. That is of course the number

$$M = \binom{20}{p}.$$

We also need to know the number of different ways to choose $m - p$ failures from among the 60 numbers that the state did not pick. Of course this number is

$$N = \binom{60}{m - p}.$$

We conclude that the product

$$M \cdot N$$

is the number of different ways to get exactly p matches with the state's chosen numbers.

We must divide this number by the total number of ways to choose m numbers from among the 80 in the grid. This number is of course

$$W = \binom{80}{m}.$$

Thus the probability of getting p hits when choosing m numbers is

$$P(p) = \frac{M \cdot N}{W}.$$

Among those who play Quick Draw regularly, a popular choice for m is 4. Thus the player chooses 4 numbers from the 80-number array. We are interested in knowing the probabilities of getting 0 hits, 1 hit, 2 hits, 3 hits, or 4 hits. It is straightforward, using a calculator, to determine that

$$P(0) = \frac{\binom{20}{0} \cdot \binom{60}{4}}{\binom{80}{4}} = \frac{1 \cdot 487635}{1581580} \approx 0.308$$

$$P(1) = \frac{\binom{20}{1} \cdot \binom{60}{3}}{\binom{80}{4}} = \frac{20 \cdot 34220}{1581580} \approx 0.433$$

$$P(2) = \frac{\binom{20}{2} \cdot \binom{60}{2}}{\binom{80}{4}} = \frac{190 \cdot 1770}{1581580} \approx 0.213$$

$$P(3) = \frac{\binom{20}{3} \cdot \binom{60}{1}}{\binom{80}{4}} = \frac{1140 \cdot 60}{1581580} \approx 0.043$$

$$P(4) = \frac{\binom{20}{4} \cdot \binom{60}{0}}{\binom{08}{4}} = \frac{4845 \cdot 1}{1581580} \approx 0.003.$$

For the 4-spot game, the state advertises on the back of the ticket that the payoff on a $1 bet is as follows:

No. spots	Payoff
0	$0
1	$0
2	$1
3	$5
4	$55

Therefore the expected payoff for a $1 bet on the 4-spot game is

$$0.308 \cdot 0 + 0.433 \cdot 0 + 0.213 \cdot 1 + 0.043 \cdot 5 + 0.003 \cdot 55 = \$0.59.$$

The obvious conclusion to draw here—that on average, if you bet \$1, you are going to make back about 60 cents—is that this is a losing game.

However, a few years ago the State of New York advertised that, every Wednesday during the month of November, it would double the payoff on the 4-spot Quick Draw game! This gives one pause for thought. That makes the expected payoff \$1.18. Which means that a \$1 bet will get you an average payout of \$1.18. In other words, you can expect to make nearly a 20% profit.

Well, that changes everything. A group of students, who had recently taken a probability course and mastered the ideas being presented here, decided to take advantage of the situation. On the first Wednesday in November, they showed up in a bar that sold Quick Draw tickets. They had calculated how many tickets they would need to buy in order to guarantee a profit. And they began to buy Quick Draw tickets by the hundreds. In fact they were purchasing more than 1000 tickets per day, and playing twenty games on each ticket. As they played more and more games, they accrued profits and were thereby able to buy *even more* tickets. The only thing that slowed them down was the speed of the printer that printed out the tickets.

By the end of November this group of students had amassed a considerable profit—in the millions of dollars. After purchasing a new house and a new car, one of the students was asked to comment on the experience. He said, "It shows that paying attention in math class can, in fact, be useful."

It is also the case that a group of students at M.I.T. beat the Massachusetts lottery. Their plan was different, as the state of Massachusetts *did not* double the winnings on any of the lottery games. These students played the Cash WinFall game. There was a quirk in the way that the jackpot was broken down into smaller prizes in case there was no big winner (as you know, in these circumstances the jackpot is rolled over to the following week). They were able to determine that a 20% profit was possible if they bought \$100,000 worth of tickets on certain days. They ultimately, over a period of years, earned \$8 million with their scheme. Curiously, the state of Massachusetts figured out what they were up to but they did nothing to stop the students. Because the state was making money too!!

Exercise: You play the following game with a friend. You work with a grid of 40 integers. You pick 10 and he/she picks 10. What is the probability that you will get 4 matches?

7.3 The Eudaemonic Pie

Not a word that you would commonly use in casual conversation is it? It refers to happiness as a result of a good life lived logically. What does this have to do with the subject of this book? Quite a lot, actually.

The story that we tell here is based on the book [BAS], which has the same name as the title of this section. The book describes a group of about two dozen members of the Physics and Mathematics Departments (both faculty and students) at the University of California at Santa Cruz, calling themselves the Chaos Cabal, who set out to beat the roulette wheels in Las Vegas. The author Thomas Bass of [BAS] was actually a member of the Cabal.

What is roulette? It is a game of chance, invented in 18th-century France. Refer to Figure 7.5 for the appearance of a roulette wheel. There is a small steel ball, and the ball is set in motion by the toss of the croupier, a casino employee, who sends the ball spinning around the wheel a few dozen times and sets the roulette wheel spinning rapidly in the opposite direction. As the ball's momentum dies, it bounces off the opposite-spinning rotor several times before settling in a particular numbered slot. The slots are numbered 1 through 36 as well as 0 and 00. So there are 38 possible slots for the ball to land in.[1]

Meanwhile, the players place their money on various numbers on a felt layout of the numbers 0, 00, and 1–36. The odds against any particular number coming up are obviously 1 in 38. But the house only pays 36 to 1. So the house has a built-in guarantee of about a 5% profit. An unusual feature of the game is that the house allows betting to go on even while the steel ball is caroming around the roulette wheel (before it starts to settle into a slot). It is here (at least in part) where the Chaos Cabal went to work.

The Chaos Cabal included a number of talented and remarkable people, including Doyne Farmer, Jim Crutchfield, and Norman Packard. These people went on in later life to founding the Los Alamos National Laboratory Center for Nonlinear Studies and the Santa Fe Institute for the Study of Complex Systems. They spent six years perfecting their system for analyzing the game of roulette and figuring out how to beat the wheel.

---

[1]In the old days a roulette wheel had only the numbers 1 through 36. The numbers 0 and 00 were added to augment the house's odds. See the next paragraph.

Figure 7.5: A roulette wheel.

The Chaos Cabal is to be admired for their creativity and determination. What they did is to design and program a computer that could be hidden in a shoe. The state of computer hardware in the year 1976 was primitive indeed. The nature and quality of memory chips, input-output devices, CPU speed, and other factors were truly inhibiting. And getting the whole computer to fit into the heel of a shoe was an immense feat.

It may be noted that Santa Cruz (the alma mater of this author) has always been an unusual place. For many years the place had no grades. Instead the professors were supposed to write a little essay about each student. And all courses were 5 credits, so that you typically only took 3 courses per term—so that you could spend more time on each course. And there were a *great many* very unusual courses.

How did this cabal of academics propose to beat the wheel in Las Vegas? A good question. Their premise was that no roulette wheel (see Figure 7.5) is perfect. It will have certain imperfections and imbalances that cause certain numbers to come up more often than others (see our detailed description of the operation of a roulette wheel above). If one can calculate those imperfections, then one can take advantage of them.

But how to do that? Enter the computer. We must bear in mind that this was 1976, several years before Steve Jobs and Steve Wozniak of the Home-Brew Computer Club invented the personal computer, which they called the

Apple. Nonetheless, the germ of Silicon Valley existed (Fairchild Electronics, for instance) and computer parts could be had. They designed a computer that could be hidden in the heel of a shoe.

The Chaos Cabal had teams of two people. One person was the "observer" and the other was the bettor (the one who places the bets). The observer takes in data on which numbers come up most frequently on the given roulette wheel. It is he/she who wears the computer in his/her shoe. He/she enters data into his/her shoe computer by tapping the foot. The data is then conveyed by radio to the bettor, and the bettor then knows what bets to make.

It turns out that the Cabal system actually gave them a 44% edge against the house. The potential to make money was really there! And they actually had investors whose money they could risk in the scheme.

There are three things that went wrong with the system developed by the Chaos Cabal:

- The rate at which they made money was not fast enough for the investors, so the investors eventually evaporated.

- The circuitry of the "eye in the sky" cameras that every casino uses to monitor dishonest croupiers and cheating customers interfered with the radio circuitry of the shoe computers.

- The members of the teams became so nervous that they sweated copiously and shorted out their electronics.

Of course casinos do not like people who try to beat the system using extracurricular means. If the Chaos Cabal people had been caught, then at the least they would have been banned from the casino. In the worst case scenario they would have been dragged into the back room and "worked over" (this would have included verbal and psychological harassment; the sort of physical harassment that you saw in Robert De Niro's movie *Casino* seems to have gone by the wayside).

There are many stories of techies and math people cooking up schemes to beat the casinos. One of the most famous ones is that of Edward Thorp who, in 1962, wrote the book *Beat the Dealer* [THOR1]. This volume told of a very sophisticated card-counting scheme that allowed one to win at the card game blackjack. And it really worked! People bought copies of this book by the carload and rushed off to Vegas to ply their wares. Well, the people who run the casinos are no fools, and they made changes in their operation to fight the success of the card counters. For example, if you go to Las Vegas today to play blackjack you will find that the dealer never works with a single deck of 52 cards. Instead he/she has a shoe with eight decks in it. Also the dealer will sometimes throw a few cards on the floor so that the players do not know which cards are missing!

A few years later Thorp, along with finance professor Sheen T. Kassouf, wrote a second book called *Beat the Market* [THOR2]. This was a book about how to apply some of the same principles as in the first book in order to win at the stock market. In fact these strategies were so successful that Thorp ended up opening his own investment house, and today he is a billionaire.

Exercise: The game of blackjack is played this way. You play with a standard deck of 52 playing cards. Cards 1 through 10 count their face value. The face cards each count 10. The ace can count as either 1 or 11. Your goal is to hold cards that sum to as close to 21 as you can get—without going over 21. You begin with 2 cards. If have a face card and an ace then that is blackjack—the best possible hand. If you have a modest hand—with 2 cards that sum to less than 21, then you may request another card. Or you may not. You have to consider the odds of going *over* 21—in which case you would lose. You are playing against the dealer.

The dealer plays by a hard and fast rule. He she begins with 2 cards just like you do. If the sum of his/her 2 cards is 16 or less, then he/she takes another card. If the sum of his/her 2 cards is 17 or more, then he/she takes no more cards. Explain in probabilistic terms why the cutoff here is 16/17.

7.4   A Dicey Bet

You hand a friend a standard deck of 52 playing cards face down. You ask him to divide the deck into three sub-decks, using simple cuts, and to place them face down on the table. Then you say "I'll bet you even money that one of those three top cards is a face card" (here a face card is a jack, queen, or king). See Figure 7.6.

Figure 7.6: Three subdecks of cards.

Would your friend be wise to accept the bet?

Your friend may be thinking that there are only 12 face cards in a deck of 52. The chances of selecting a face card are therefore $12/52 \approx .2308$. Clearly the bet is in his favor and he should accept it.

Unfortunately, your friend (if he is indeed thinking this way) does not understand the concept of sample space, nor how to count accurately. Here is a correct analysis.

If you set aside all the hocus pocus, then what you and your friend are doing is selecting three cards at random from a deck of 52. The question is what is the probability that one of those three cards is a face card. Now there are $\binom{52}{3}$ ways to select three cards from among 52 (see Section 7.1). It turns out—and this is true in many probability problems—to be convenient to now calculate how many ways there are of *not* choosing a face card. In other words, we shall calculate the probability of failing, rather than succeeding.

If we are going to select three cards, none of which is a face card, then we must select three cards from among the $40 = 52 - 12$ that are not face cards. The number of ways of doing this is $\binom{40}{3}$. Thus the probability of our *failing* to select a face card is

$$
\begin{aligned}
\frac{\binom{40}{3}}{\binom{52}{3}} &= \frac{40!/(3! \cdot 37!)}{52!/(3! \cdot 49!)} \\
&= \frac{49 \cdot 48 \cdot 47 \cdot 46 \cdot 45 \cdot 44 \cdot 43 \cdot 42 \cdot 41 \cdot 40 \cdot 39 \cdot 38}{52 \cdot 51 \cdot 50 \cdot 49 \cdot 48 \cdot 47 \cdot 46 \cdot 45 \cdot 44 \cdot 43 \cdot 42 \cdot 41} \\
&= \frac{40 \cdot 39 \cdot 38}{52 \cdot 51 \cdot 50} \\
&\approx .44706.
\end{aligned}
$$

As a result of this calculation, the probability that we *did* get a face card among the three that we selected is

$$P = 1 - .44706 = .55294.$$

Notice that the odds are *better than even* that one of the three top cards is a face card. Therefore the situation described is a good bet for you and not a good bet for your friend.

Exercise: Suppose you modify the game by instead dividing the deck into *four* subdecks. *Now* what is the probability that one of the top cards is a face card?

7.5   The Game of Life

The Game of Life (invented by John Horton Conway) is played on a board divided into squares—like a large piece of graph paper. Begin by placing x's on some of the squares. These are the "people" in your population. Two people (squares) are "neighbors" if they share either an edge or a corner. Thus each square has eight neighbors—four to the left and right or above and below, and four diagonal.

The rules of the game are these: (i) if three people are neighbors to the same empty square, then they produce an offspring (another person) in that square, (ii) if some person has four or more neighbors, then it dies of overcrowding, (iii) if some person has one or fewer neighbors, then it dies of loneliness. Given any population configuration, all three rules are applied instantly to produce the next configuration.

Is there any initial population that will produce a population that is periodic—i.e., develops a pattern that, after finitely many steps, keeps repeating (try three squares in a horizontal row)? Is there any initial population that will remain static and never increase or decrease? Is there any initial population that will die out quickly—or immediately? Is there any that will keep reproducing and become ever larger without bound?

A configuration that produces a periodic population is shown in Figure 7.7.

Unless we have infinitely many people, a static configuration is impossible. If there are only one or two squares, they die. If there are more, one can

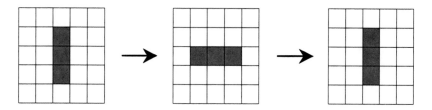

Figure 7.7: A periodic population.

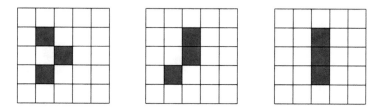

Figure 7.8: Production of offspring.

easily check that if one person has three neighbors, an offspring is always produced. Thus, for the configuration to remain static, each person must have at most two neighbors. On the other hand, the configurations in Figure 7.8 also produce offspring.

Thus, the only possibility is if all the squares are arranged in a diagonal line—see Figure 7.9.

But then the ones at the ends will die of loneliness (of course, unless the line has no ends, for which we need infinitely many people).

A configuration that dies immediately is shown in Figure 7.10.

There are configurations that grow without bound. One such configuration was found in 1970 by B. W. Gosper. It is called the "Glider Gun." It is shown in Figure 7.11.

The reader can check that every 30 generations, a new "glide" is shot from the gun (top of the figure) in the southeast direction. For more on "The Game of Life" and other games, we refer the interested reader to the book *Winning Ways for Your Mathematical Plays* by Elwyn R. Berlekamp, John H. Conway, and Richard K. Guy [BCG].

Exercise: Find a configuration for the game of life that will die out in 3 moves. Find one that will die out in 5 moves.

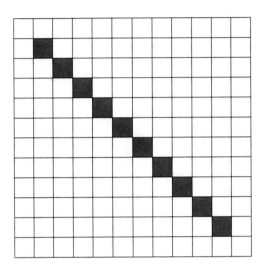

Figure 7.9: A static configuration.

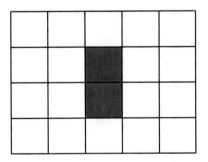

Figure 7.10: A configuration that dies immediately.

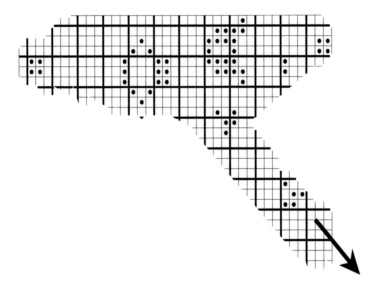

Figure 7.11: The Glider Gun.

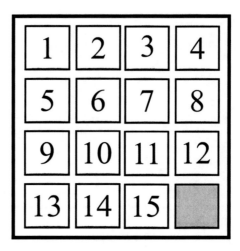

Figure 7.12: Sam Loyd's puzzle.

7.6   The Tower of Hanoi

The Tower of Hanoi is a puzzle that was created in 1883 by the French mathematician Édouard Lucas. That was an age when puzzles and puzzle-solving were enjoying a popular rage. Puzzles appeared almost daily in the newspapers, and books of puzzles sold widely. Sam Loyd was perhaps the most famous and successful puzzle-maker of the day. His *Sam Loyd's Puzzle* sold millions of copies (see Figure 7.12), and Loyd made a good living creating and selling puzzles. See the book [LOY] for a sample of his work.

What makes the Tower of Hanoi particularly interesting is that is not merely a device for whiling away the time. In fact, many of the most incisive schemes for backing up computer systems are modeled on the Tower of Hanoi. The Tower of Hanoi is used as a test by neuropsychologists attempting to evaluate frontal lobe deficits. It is used in psychological research in problem solving. It is also used for neuropsychological diagnosis and treatment of executive functions.

Figure 7.13 shows what the Tower of Hanoi looks like. As you can see, it consists of a stack of discs, having descending size as you go up, centered on one of three spindles.

The challenge is to move the discs to another spindle, while following certain rules. The rules are these:

Figure 7.13: The Tower of Hanoi.

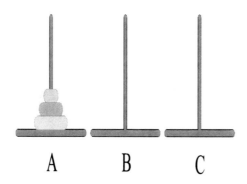

A   B   C

Figure 7.14: The Tower of Hanoi with three discs and three spindles.

(a) You can only move one disc at a time.

(b) Each move consists of taking one disc from the top of a stack and placing it on top of another stack.

(c) You can only set a disc on top of a disc that is larger (not smaller) than itself.

This problem has been studied extensively, and there are a number of scholarly articles on the topic. It is known that, in order to solve the problem with n discs, it takes about $2^n - 1$ moves. We shall illustrate that idea now with a stack of three discs placed on the first spindle. In fact we shall suppose that the spindles are labeled, left-to-right, as A, B, C. See Figure 5.14.

Let us number the discs 1, 2, 3, with 1 the smallest (on top) and 3 the largest (on the bottom). For the first move, we must lift disc 1 and place it somewhere else. Let us put disc 1 on spindle C. Now, for move two, we shall do something with disc 2. We cannot place it atop disc 1, so instead we put disc 2 on spindle B. Now move three is more interesting. We cannot move disc 3 anywhere. If we want to move disc 2 the only legal move is to put it back where it came from on spindle A. That makes no sense. So we must move disc 1, and we shall put it on top of disc 2 on spindle B. So far we have made three moves.

For move four, we shall place disc 3 on spindle C. Then, for move five, we place disc 1 on spindle A. Now, for move six, we put disc 2 on top of disc 3 on spindle C. Finally, for move seven, we lift disc 1 off spindle A and put it on top of the discs on spindle C.

We see that, in seven moves (and $7 = 2^3 - 1$, as predicted), we have transferred the three discs from spindle 1 to spindle 3. That is a successful solution to the Tower of Hanoi problem for three discs.

There is in fact an algorithm for solving the Tower of Hanoi. It goes like this (in two cases):

The Case of an Even Number of Discs

1. Make the legal move between spindles A and B.

2. Make the legal move between spindles A and C.

3. Make the legal move between spindles B and C.

4. Repeat these steps until the process is complete.

The Case of an Odd Number of Discs

1. Make the legal move between spindles A and C.

2. Make the legal move between spindles A and B.

3. Make the legal move between spindles C and B.

4. Repeat these steps until the process is complete.

You may wish to try these with three discs and with four discs.

Exercise: In fact there is a single algorithm that works all the time. We summarize it now:

1. Take the disc with the least number attached to it and move it as far to the right as possible.

2. Now take the next least numbered disc and move it as far to the right as possible.

3. Continue in this fashion, always obeying the three basic rules of the Tower of Hanoi.

Try this unified algorithm on our original example of three discs. You will find that it duplicates exactly the solution that we already presented.

Exercise: Imagine a Tower of Hanoi puzzle in which there are two separate stacks of five discs. And your job is to switch the two stacks. Can you do this with just three spindles? How about four spindles? How many spindles does it take to solve this puzzle? Is it clear that the problem can be solved at all?

Chapter 8

Geometry

8.1 Thoughts of Pythagoras

The Pythagoreans, who lived about 2500 years ago, were both a scientific society and a religious one. They lived as a colony, and generated many innovative ideas.

Certainly one of the most daring and influential ideas to come from the Pythagoreans was the notion that not all numbers are rational. In fact the Greeks had previously thought of all numbers as the ratios of integer sides of a right triangle. See Figure 8.1.

The Pythagoreans addressed this (mistaken) belief by doing two dramatic things. First they proved the notable *Pythagorean theorem*. This says that the two legs a and b of a right triangle are related to the hypotenuse c by the formula

$$a^2 + b^2 = c^2 .$$

See Figure 8.2. Let us see now why this is true.[1]

Examine Figure 8.3.

We see a square that contains four identical triangles (in fact these are all identical to the original Pythagorean triangle in Figure 8.2) and a smaller

---

[1]In fact, this Pythagorean theorem has more proofs than any other result in mathematics. One of the proofs was created by American President Grover Cleveland. We present here one of the simplest and most direct proofs.

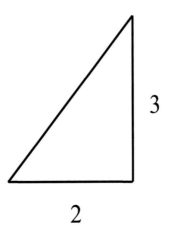

Figure 8.1: The Greek notion of number.

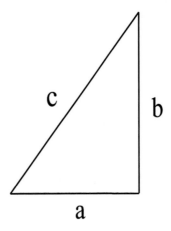

Figure 8.2: The Pythagorean theorem.

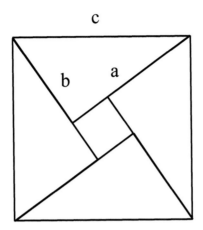

Figure 8.3: Proof of the Pythagorean theorem.

square. Now the big square obviously has area $c \times c = c^2$. Each triangle is a right triangle with legs a and b; hence each has area $ab/2$. And the smaller square has side $(a - b)$ and hence area $(a - b)^2$.

Of course the sum of the areas of the four triangles and the smaller square equals the area of the larger square. We conclude that

$$c^2 = 4 \cdot \left(\frac{1}{2}ab \right) + (a - b)^2 .$$

Multiplying out the right-hand side, we see that

$$c^2 = 2ab + (a^2 - 2ab + b^2)$$

or

$$c^2 = a^2 + b^2 .$$

That is the Pythagorean theorem.

Now the Pythagoreans studied this result and noticed that, if the right triangle has two legs of lengths 1 and 1, then the hypotenuse c satisfies

$$c^2 = 1^2 + 1^2 = 2 .$$

So they asked themselves what sort of number c could be. And the answer is startling: c cannot be rational. Here is why.

Suppose that c is rational. Then $c = p/q$, where p and q are positive whole numbers. We may as well assume that the fraction p/q is reduced to lowest terms, so that p and q have no common factors. So, because $c^2 = 2$, we see that

$$\left(\frac{p}{q}\right)^2 = 2.$$

In other words,

$$\frac{p^2}{q^2} = 2$$

hence

$$p^2 = 2q^2.$$

Since 2 divides the right-hand side, it also must divide the left-hand side. But then we must conclude that 2 divides p. Hence $p = 2k$ for some integer k. Substituting this fact into the last equation we see that

$$(2k)^2 = 2q^2$$

so that

$$4k^2 = 2q^2$$

or

$$2k^2 = q^2.$$

Since 2 divides the left-hand side, it also must divide the right-hand side. So 2 divides q.

We have shown that 2 must divide p and also 2 must divide q. But this contradicts our initial hypothesis that p and q have no common factors. That is a contradiction. Therefore it is impossible to write c as p/q. Thus c must be irrational.

These days we understand that the real number system consists of rational numbers and irrational numbers. It is convenient to think of the real numbers as the set of all possible decimal expansions. Those decimal expansions that contain no infinite repetitions are the irrational numbers. Those expansions that contain infinite repetitions or are terminating are the rational numbers. For example,

$$7.643502$$

and

$$3.41723232323\ldots$$

Figure 8.4: A butterfly is symmetric.

Figure 8.5: A snowflake is symmetric.

are both rational numbers. By contrast, the number

$$3.14159265\ldots$$

(otherwise known as π) is an irrational number.

Exercise: Explain why between any two rational numbers there is an irrational number. Explain why between any two irrational numbers there is a rational number.

8.2 Symmetry

We all have an innate sense of what symmetry is. We think of a butterfly or a snowflake as being symmetric. See Figures 8.4 and 8.5.

We know that a circle, or disc is *very* symmetric while a square is only somewhat symmetric. See Figure 8.6.

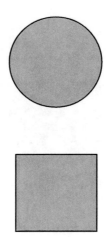

Figure 8.6: Symmetry of the disc and the square.

Figure 8.7: A region with no symmetry.

To take the thought to an extreme, look at the region in Figure 8.7. Most people would say that this region has no symmetry at all.

But these are all intuitive statements. What exactly do we mean by symmetry? This is the sort of question that a mathematician can really sink his/her teeth into.

Suppose that you and I are standing together in a room with a circular table. You leave the room and I lift up the table and rotate it a certain number of degrees—say 37° to be specific. Then I put the table down in the same place on the floor that it was before. You walk back into the room. And you cannot tell that I have rotated the table, nor can you tell through how many degrees I have rotated the table.

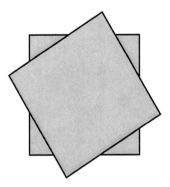

Figure 8.8: Rotation of a square table.

But if instead we are in a room with a square table, and you leave the room, and I rotate the table (again let us say through an angle of 37°) and replace the table where it stood before, then when you return to the room you will be able to tell immediately that the table has been rotated. In fact there are only four angles through which I could rotate the square table so that I can fool you (that is, you cannot tell that the table has been rotated). These are the angles 90°, 180°, 270°, and 360°. See Figure 8.8.

The rotations of the square table bear further thought. When we were thinking of rotations through the angles 90°, 180°, 270°, and 360°, we were probably thinking of *counterclockwise* rotations—because that is just how people think. But we could also rotate *clockwise* through any of these angles, and that would preserve the top of the square table. Also note that the clockwise rotation through 90° *undoes* the counterclockwise rotation through 90°. And likewise for the other three angles. So each special rotation has an inverse.

Further observe that the rotation through 360° is just the same as the rotation through 0°. Neither of these rotations moves any point of the surface of the table. Either of these rotations is like the "zero element" in a number system.

Perhaps most interesting is that we can *compose* these rotations. For example you can follow the rotation through 90° by the rotation through 180° and obtain the rotation through 270°. In Figure 8.9, on the left, you can see that the black dot on the right will be taken by this composition to the black dot on the bottom. Or you can follow the rotation through 180° by the rotation through 270° and obtain the rotation through 90°. In Figure

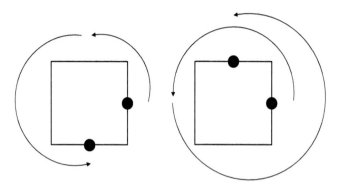

Figure 8.9: Composition of rotations.

8.9, on the right, you can see that the black dot on the right will be taken by this composition to the black dot at the top.

It is interesting to note that, in the second example in the last paragraph, we composed two rotations whose *total rotation* was 450°—more than 360°. In other words, the superposition of these two rotations was more than a full circle. But we understand that one subtracts off any multiples of 360° to obtain a rotation that is between 0° and 360°. In this case, 450° − 360° = 90°. That way we are combining two of our rotations, chosen from the list 90°, 180°, 270°, and 360°, to obtain a new element from that list. We composed 270° with 180° to obtain 90°.

The collection of symmetries on the square table forms what mathematicians call a *group*. This is an idea that goes back to the French mathematicians Evariste Galois (1811–1832) and Augustin Cauchy (1789–1857). The key concept is that you have a set that is equipped with a binary operation. This operation allows you to combine two elements of the set to produce a new element of the set. And the operation has certain significant properties: it has a zero element, and each element has an inverse element.

Contrast the situation for the square table with the situation for the circular table. The circular table has many more symmetries. Indeed, you can rotate the circular table through *any angle* between 0° and 360° and preserve the table so that one cannot tell that it has been rotated. Notice that it is still possible to compose two rotations to obtain a third rotation. Also, each rotation has an inverse (namely, the inverse of the rotation of θ degrees in a counterclockwise direction is the rotation of θ degrees in a clockwise direction). Finally, there is a "zero" rotation—the rotation that

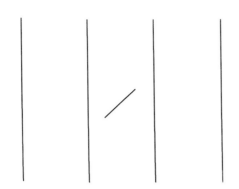

Figure 8.10: Long planks that are six inches wide.

involves no motion at all. So the rotations of the circular table also form a group. And it is a *much larger* group than the group of rotations for the square table.

Mathematics and physics are full of groups. Thanks to the work of John von Neumann (1903–1957), we now study quantum mechanics using group theory. Much of number theory and modern algebra uses group theory. Modern cryptography uses a considerable amount of group theory. Algebraic structures in general (rings, fields, modules, etc.) are a big part of today's mathematics. This section has given you just a taste of these ideas.

Exercise: Consider the unit cube U in three-dimensional space. The set of rigid motions of space that preserve U form a group. Explain why this is so. And calculate all the elements of the group. How many group elements are there?

8.3   Buffon's Needle Problem

The problem discussed here was originally posed by Georges-Louis Leclerc, Comte de Buffon. A floor is made of long planks that are six inches wide (Figure 8.10).

A girl drops a thin stick that is four inches long onto the floor. She does so a great many times—say N times. Let us stress that the stick is dropped in a completely random manner—so that sometimes it will cross a crack

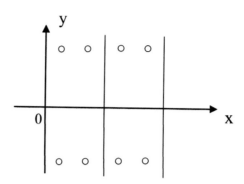

Figure 8.11: A coordinate system for Buffon.

between the planks and sometimes it will not. Calculate the probability that the stick will land on a crack (between two planks) in the floor as a function of N. What happens to this probability as $N \to \infty$?

This is a classic problem that is known as Buffon's Needle Problem. Clearly the probability that the stick will land on a crack depends on the angle at which the stick falls. For instance, if the stick falls parallel to the cracks, then it is highly unlikely that the stick will cross a crack. If, instead, the stick is perpendicular to the cracks, then it is fairly likely that it will hit a crack. Since the question of whether the stick will cross a crack depends on the angle at which the stick falls, you may not be surprised to learn that our answer will involve π. In fact, it turns out that dropping a stick on the floor is one method for calculating π.

We set up a coordinate system as in Figure 8.11. The x-axis is perpendicular to the direction of the cracks. The y-axis is parallel to the cracks. We arrange for the origin to lie on a crack, so that the y-axis runs along a crack.

In order to make this problem tractable, we are now going to make some normalizing assumptions. First, we assume that the stick is infinitely thin— like a line segment. Next, we assume that one end of the stick is painted red. We measure the "angle of the stick" as follows: translate the stick, without rotating it, so that the unpainted end is at the origin. Now measure the directed angle beginning at the positive x-axis, in the counterclockwise direction to the stick (just as you do when learning about the sine and cosine functions). If that angle is θ radians, then we say that the stick subtends an angle of θ radians with the positive x-axis.

Now fix an angle $0 \leq \theta \leq \pi/2$. We restrict attention to a stick that will fall onto the floor in such a fashion that it subtends an angle of θ, but is otherwise random. What is the probability that it will cross a crack? Clearly the vertical position of the stick is of no interest, for it does not affect whether the stick crosses a crack. All that matters is the left-right position of the stick. And the problem is periodic: As the left end of the stick moves from 0 to 6, with the angle staying the same, the following occurs. For a certain distance the stick will not touch a crack. Then it starts to cross a crack and the stick will move left-to-right across the crack. When the left end of the stick reaches 6, we are again at a situation that is identical to when the left end of the stick is at 0.

Now we use a little trigonometry. The stick has length 4 inches. When it subtends angle θ, then the left-to-right extent of the stick is $4\cos\theta$. Thus when the left end of the stick is between 0 and $6 - 4\cos\theta$, then the stick will not cross a crack. However when the left end of the stick is between $6 - 4\cos\theta$ and 6, then the stick *will* cross a crack. Thus the probability that a stick subtending an angle θ will cross a crack is

$$P_\theta = \frac{6 - (6 - 4\cos\theta)}{6} = \frac{4\cos\theta}{6}.$$

A moment's thought reveals that the situation repeats itself when $\pi/2 \leq \theta \leq \pi$ or when $\pi \leq \theta \leq 3\pi/2$ or when $3\pi/2 \leq \theta \leq 2\pi$. So we may concentrate attention on $0 \leq \theta \leq \pi/2$.

The stick is just as likely to land so that it subtends any one angle as it is to land so that it subtends any other angle. Thus all the probabilities P_θ are equally likely to apply. We obtain our final answer by averaging P_θ over $0 \leq \theta \leq \pi/2$. Thus the ultimate probability that our stick will cross a crack is

$$\frac{1}{\pi/2} \int_0^{\pi/2} \frac{4\cos\theta}{6}\, d\theta = \frac{4}{3\pi} \int_0^{\pi/2} \cos\theta\, d\theta = \frac{4}{3\pi}. \qquad \square$$

[*Hint:* This is the only place in the solution of the present problem that we actually use calculus. And it could be avoided. Indeed, it is avoided by using the ideas that go into developing calculus. Instead of averaging P_θ using an integral—something which you may have never seen—instead do this. Divide the interval $[0, \pi/2]$ into one hundred tiny subintervals of equal length. Evaluate P_θ at one hundred values of θ, one chosen from among each of these subintervals. (Use a computer, of course.) Add together the one hundred values that you obtain and divide by 100. You will obtain an answer that is very close to $4/(3\pi)$.]

The author of this book performed an experiment dropping a stick of length 4 inches onto a floor with 6 inch wide planks. After 100 drops, the stick crossed a crack 46 times. Thus the calculated probability of hitting a crack is .46. Equating this with the expression $4/(3\pi)$ yields an approximate value for π of 2.9. This is not very accurate, and one would expect to have to perform perhaps a great many more drops to achieve any real accuracy.

In 1850, an experiment was performed by Wolf in Zurich. He used a needle of length 36 mm and planking of width 45 mm. Of course he had to adjust the formula from the last problem accordingly. After dropping the needle 5000 times he arrived at the value 3.1596 for π. Analogous experiments conducted by Fox in England in 1864, using 1100 throws, gave a value of 3.1419 for π. Smith in England (1855) used 3200 throws and found the value 3.1553 for π.

Exercise: Do an analysis of Buffon's Needle Problem when the planks are distance d apart and the stick has length ℓ. What interesting new feature appears when $\ell > d$?

8.4   Euler's Formula

Suppose that we have an admissible graph on the unit sphere in three-dimensional space. Here, by "admissible graph" we mean a connected configuration of arcs. Two arcs may be joined only at their endpoints. The endpoints of the arcs in the graph are called *vertices*. The arcs are called *edges*. An edge is that portion of an arc that lies between two vertices. A *face* is any two-dimensional region, without holes, that is bordered by edges and vertices. Figure 8.12 illustrates an admissible graph and a non-admissible graph. The graph on the left has 1 face, 4 edges, and 4 vertices.

This problem asks you to verify *Euler's formula* for an admissible graph. We let V be the number of vertices, E the number of edges, and F the

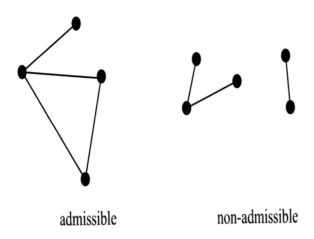

admissible   non-admissible

Figure 8.12: An admissible graph and a non-admissible graph.

number of faces. Then Euler's formula is

$$V - E + F = 2.$$

We begin with some special cases, just to be sure that we understand what is going on.

The simplest graph that is admissible, according to our definitions, consists of a single vertex and nothing else (Figure 8.13). The complement of that single vertex, in the sphere, is a valid face. Thus $V = 1$, $E = 0$, and $F = 1$. Then

$$V - E + F = 1 - 0 + 1 = 2$$

and we see that Euler's formula is valid.

The next most complex graph has one edge, with a vertex on each end, and nothing else. The complement (in the sphere) of this edge with its endpoints is a single valid face. See Figure 8.14. Thus, in this case, $V = 2$, $E = 1$, and $F = 1$. We see that

$$V - E + F = 2 - 1 + 1 = 2.$$

Thus Euler's formula is valid in this case as well.

Now we let $P(k)$ be the statement "Euler's formula is valid for any admissible graph with k edges." We shall use induction to prove this statement for every k.

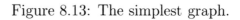

Figure 8.13: The simplest graph.

Figure 8.14: The next most complex graph.

The statement $P(1)$ has already been verified. That is part (i) of the method of induction.

For part (ii), we assume that Euler's formula is valid for any admissible graph having j edges. Now let $\mathcal{G}$ be a graph having $(j + 1)$ edges. There is some edge that can be removed from $\mathcal{G}$ so that the remaining graph $\mathcal{G}'$ is still admissible (exercise—for example, an edge that separates two different faces will do). Say that V', E', and F' denote the numbers of vertices, edges and faces for the graph $\mathcal{G}'$. Now consider what the corresponding numbers V, E, F for the graph $\mathcal{G}$ might be.

The graph $\mathcal{G}$ is obtained from $\mathcal{G}'$ (we are reversing the construction that produced $\mathcal{G}'$) by adding an edge. If the edge is added by attaching *one end* and leaving the other free (the added edge is dotted in Figure 8.15), then the number of faces does not change, the number of edges is increased by one, and the number of vertices is increased by one. See Figure 8.25. Thus $V = V' + 1$, $E = D' + 1$, and $F = F'$. Since, by hypothesis, $V' - E' + F' = 2$, it follows that $V - E + F = 2$ as desired.

If instead the edge is added by attaching *both ends* (again the added edge is dotted in Figure 8.16—there are two possibilities, as shown), then the number of faces is increased by one, the number of edges is increased by one, and the number of vertices does not change. Therefore $V = V'$, $E = E' + 1$, and $F = F' + 1$. Since, by hypothesis, $V' - E' + F' = 2$, it follows that $V - E + F = 2$.

Since these are the only two ways that a new edge can be attached, we have established step (ii) of the inductive procedure. Our argument is complete.

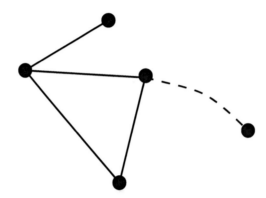

Figure 8.15: Adding an edge by means of one vertex.

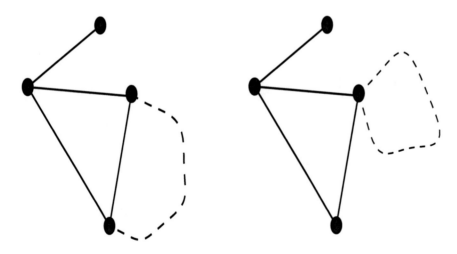

Figure 8.16: Adding an edge in two different ways.

Figure 8.17: The rectilinear circle packing.

Exercise: What does Euler's formula say when you consider graphs on the torus? [**Hint:** The left-hand side will still be $V - E + F$ but the right-hand side will no longer be 2.]

8.5   Sphere Packing

It is an interesting question, with roots in the famous list of 23 problems that David Hilbert promulgated in 1900 to guide mathematical research in the twentieth century (Hilbert's problem 18 in particular), to determine how most efficiently to pack pennies in the plane or pack balls in space.

The most obvious way to pack pennies is the so-called *rectilinear packing*, which is illustrated in Figure 8.17.

It is elementary to calculate that, with the rectilinear packing, the percentage of the plane that is covered by pennies (as opposed to the percentage that is *uncovered*) is about 78.5%. Note that 22.5% is not covered. Is that the best percentage that can be achieved? You may be surprised to learn that the answer is "no." In fact, it can be proved that the optimal packing

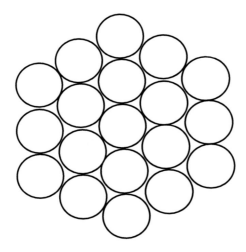

Figure 8.18: The hexagonal circle packing.

by pennies is the *hexagonal packing*. This packing is illustrated in Figure 8.18. Now it may be calculated that the percentage of the plane covered by pennies is about 91%. That is a considerable improvement.

It was an open problem for many years to determine the optimal backing of balls in space. Just to be clear, we assume that all the balls have the same size and (spherical) shape. Our job is to stack them in the fashion that will cause the greatest percentage of space to be filled up by balls.

This three-dimensional problem in fact has an interesting history. The question was first formulated by Sir Walter Raleigh, the man who brought tobacco to America. In addition to his social and political functions, Raleigh was an adventurer and a sailor and something of a pirate. One day, aboard his ship, he asked his gunman what was the most efficient way to stack cannonballs. A firm answer to the question was inaccessible at the time, and the problem (in mathematical form) ultimately made its way to the astronomer/mathematician Johannes Kepler. Kepler was a very distinguished figure, having made a name for himself by formulating three laws for the motions of the planets about the sun (see Section 5.4). The classical formulation of Kepler's problem is then, "What is the most efficient way to pack balls of the same size into space?" Kepler published this question in a booklet called *Strena sue de nive sexangula* in 1611. Interesting. Of course he was unable to answer the question, nor even to begin to answer it.

It is and was generally believed that the 3-dimensional analogue of the planar hexagonal packing is the optimal packing for spheres in three dimensions (refer to Figure 8.19). This is the "default" manner in which a grocer would stack oranges in a store display, or a gunner would stack his cannonballs. It has density $\alpha = 0.74048$. That is to say, with the described packing, 74.048 percent of space will be occupied by the cannonballs and 25.952 percent of space will be occupied by air. Old Mother Nature, with the way that she packs atoms into molecules, provides evidence for this belief. But there was, until recently, no proof.

Of course Kepler lived from 1571–1630—quite some time ago. The Kepler problem lingered for hundreds of years. The great mathematician Carl Friedrich Gauss (1777–1855) was able to solve the problem if it is assumed in advance that the packing of the balls forms a regular pattern or lattice. The redoubtable Buckminster Fuller (1895–1983) claimed in 1975 to have a proof of the Kepler conjecture, but he was in error.

László Fejes Toth (see [FTO]) made a name for himself in the 1950s by proving some fundamental results about the question. He wrote many papers on the Kepler problem, but one of the important upshots of his work is that he showed, using George Dantzig's theory of linear programming, that the problem can be reduced to the checking of finitely many (but a very *large* finite number) of cases. So the problem could, in principle, be attacked with a computer.

And, in fact, in the 1990s that is exactly what happened. Thomas Hales announced his result in 1998. He actually proved that the optimal configuration of balls in 3-dimensional space is the natural generalization of the hexagonal packing of pennies in the plane. In other words, it is the way that your grocer stacks oranges in the grocery store (see Figure 8.19).

Of course the computer calculations that Hales used were immense, using thousands of hours of computer time. It is impossible for any human being to check or to understand such a "proof." But there are other computer techniques for *checking* a computer proof. Hales launched a project called `Flyspeck` to check his proof (see [HAL3]). And that process was completed, thanks to the combined efforts of many scientists around the world, in 2014. So now Hales's solution of the Kepler sphere-packing problem has been validated—at least by another computer.

Interesting ideas about the problem in higher dimensions were offered by John Horton Conway of Princeton in [CGS]. Wu-Yi Hsiang proposed to use a new implementation of spherical trigonometry—in fact he virtually reinvented the subject—to give a resolution of Kepler's problem in dimension three. In 1993 he wrote a long paper [HSI1] (92 pages), together with a secondary paper [HSI2] explaining what was going on in the first paper, laying out his solution of the Kepler sphere-packing problem. This caused quite a sensation, since there had been no serious attack on the problem for a great many years. The paper [HSI3] has further developments of this problem.

But it was not long before the experts began to cast aspersions on Hsiang's efforts. It seems that Hsiang had declared that "Thus and such is the worst possible configuration of balls and we shall content ourselves with examining this configuration." Unfortunately, Hsiang did not provide a convincing reduction to the indicated special case. In fact, it is generally believed that problems like this *do not have* a worst-case scenario. One has to come up with arguments that address all cases at once. One of the leaders of the anti-Hsiang movement was Thomas Hales of the University of Michigan (now of the University of Pittsburgh). He published a polite but detailed article [HAL1] in the *Mathematical Intelligencer* taking Hsiang's efforts to task. It should be noted that others, including John Horton Conway, had endeavored to get Hsiang to admit to his errors. Conway published an important book [CGS] in the subject area and made no reference to Hsiang's work. But it was Hales who bit the bullet and made public statements to the effect that Hsiang was wrong. Hsiang replied to Hales's allegations in [HSI4].

Thomas Hales was not just blowing smoke. He had *his own* proof of the Kepler sphere-packing problem which he had intended to publish as a long sequence of papers. He studied the work of L. Fejes Tóth and concluded that the problem could be solved by minimizing a function of 150 variables. This in turn would entail solving about 100,000 linear programming problems.[2] In 1992, assisted by his graduate student Samuel P. Ferguson, Hales embarked on a program to prove the celebrated conjecture. They produced a total of six papers along the way. The final piece of the puzzle appears in the paper [HAL2], which has a colorful history.

For the record, the sphere packings in three dimensions that do the job

---

[2]Linear programming is a subject developed by George Dantzig in the late 1940s. It is used to find extrema for systems of linear equations. The methodology has far-reaching applications in scheduling airlines, routing Internet messages, and doing Google searches.

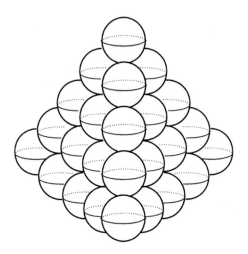

Figure 8.19: The optimal packing of spheres.

are the cubic close packing (face-centered cubic) and hexagonal close packing arrangements. The cubic close packing is exhibited in Figure 8.19. The Web page

\https://en.wikipedia.org/wiki/Close-packing_of_equal_spheres

has a nice discussion of these packings together with detailed graphics.

Hales has received considerable praise and encomia for this work. It is quite an achievement. But it leaves open many other "packing" questions.

Hales's capstone paper, in manuscript form, was more than 200 pages long. He submitted it to the *Annals of Mathematics*, which is the "journal of record" for modern mathematics. It is arguably the most prestigious mathematics journal in the world. Being a journal of such high rank, the *Annals* has rather strict rules. Generally speaking, it does not consider exceedingly long papers. It wants to consider only modern problems of current interest. And it wants to consider only papers that are the "last word" in a subject. Finally, it usually will consider only papers that have a traditional mathematical form. That is to say, the *Annals of Mathematics* has been a bastion of the classical form of mathematical proof: keenly crafted sequences of statements connected together by the strict rules of logic. Thomas Hales's paper, whatever its merits may have been, did not conform to that standard. It relied heavily on computer calculation (much like the proof of the four-color theorem from 25+ years before—see Section 2.3). But Robert MacPherson, Managing Editor of the *Annals*, was somewhat enamored of computer proofs. He liked the idea of a computer proof of this crusty old problem. And he was willing to consider it. And, after all, Hales's paper was indeed the last word in the subject.

But who could referee such a work? MacPherson was able to use the stature and clout of the *Annals* to recruit a team of twenty Hungarian mathematicians, led by Gábor Fejes Tóth (son of László Fejes Tóth, the previously mentioned pioneer in the study of the Kepler problem), to study the paper and render a judgment. They spent several years poring over the details of this long and tedious mathematical work. In fact the refereeing process was so protracted and so arduous that there was attrition among the referees: some quit and others retired or died. In the end they said that they were able to confirm the mathematical parts of the paper but it was impossible to check the computer work. Thus no conclusive report could be rendered. Actually, there is a bit more to it than that.

The twenty Hungarians ran a seminar for three semesters on the paper. They studied it for four years altogether. In the end they said they were 99% sure it was right.

But MacPherson was not daunted. He accepted Hales's paper for the *Annals of Mathematics*—it is [HAL2]. In fact what MacPherson wrote to Hales, on acceptance of the paper, was this:

> The referees put a lot of energy into this that is, in my experience, unprecedented. They ran a seminar on it for a long time. A number of people were involved, and they worked hard. They checked many local statements in the proof, and each time they found that what you claimed was in fact correct. Some of these local checks were highly non-obvious at first, and required weeks to see that they worked out. ... They have not been able to certify the correctness of the proof, and will not be able to certify it in the future, because they have run out of energy to devote to the problem.

MacPherson originally planned to publish the paper with a displayed disclaimer that said, in effect,

> The *Annals* cannot be certain that this paper is correct. Never-

theless we feel that the paper is worthwhile.

This was a bellwether for modern mathematics! Think about the long tradition of proofs in mathematics. Think about what proofs represent. This is what distinguishes mathematics from biology and physics and engineering. We do not perform experiments and come to plausible conclusions (which could be, and often are, refuted later). We instead prove theorems once and for all. Once a theorem is proved, and its proof checked and validated, then the theorem stands forever. It is just as true, and just as useful, today as when it was proved. Thus we use the Pythagorean theorem with confidence today, because Pythagoras proved it 2500 years ago. We use the prime number theorem with never a doubt today because Hadamard and de la Vallée Poussin proved it 100 years ago. The *Annals* had always taken a hard line in defense of the traditional notion of mathematical theorems and proofs. It had always had its papers refereed assiduously. The *Annals* wanted to maintain its position as the journal of record for major advances in mathematics. It wanted to be *certain* that the papers it published were correct. There is rarely a published correction or retraction of a paper published in the *Annals of Mathematics*. But now it was deviating from that line. The venerable *Annals of Mathematics* was going to publish a result that nobody could confirm!! And they were going to acknowledge this shortfall with a *disclaimer*.

When John Horton Conway found out the plans for Hales's paper in the *Annals*, he was quite perturbed. He phoned MacPherson and gave him a piece of his mind. Conway's view was that the *Annals* should not publish a paper with a disclaimer. The appellation *Annals* suggests a certain augustness and timelessness to the journal and its contents. This journal should be a showcase for mathematics that has been certified to be correct. Nothing else. He convinced MacPherson to remove the disclaimer. It has now been replaced by a ringing endorsement.[3] Note that, according to the Web page of the *Annals of Mathematics*, the paper was received on September 4, 1998, and was finally accepted on August 16, 2005. It is well known that some journals can be rather slow. One and a half to two years may be a typical time for a journal to get a new paper refereed, typeset, and in print. The slower journals may take three years or more. But seven years is quite extraordinary. It must

---

[3]At least that was the original plan. The paper [HAL2] has now appeared, and it contains *neither* a disclaimer *nor* a ringing endorsement.

be borne in mind that over four years of this time was for the refereeing!

The current version of the truth is that the *Annals* has published an *outline* of Hales's proof.[4] Entitled "A Proof of the Kepler Conjecture," this outline is 121 pages long (see [HAL2]). It is in the November, 2005 issue. The full details appeared elsewhere, in the July 2006 issue of *Discrete and Computational Geometry*. Hales's work will take up the entire issue, and will be divided into six papers. These will comprise 265 pages.

This is also new territory for the estimable *Annals*. The *Annals of Mathematics* is, and has been, a stodgy old girl. She publishes complete, self-contained papers with complete proofs of new and important results. Its publication of Thomas Hales's work is charting new ground. And it is setting an example for other journals. It is possible that the entire nature of the publication of mathematical research will be affected by these actions.[5]

Those of a more technical bent are concerned about the Hales resolution of the Kepler problem because it uses `CPlex` to solve the relevant linear programming problems. The trouble with the software `CPlex` is that, while it is reliable, it does not certify any digit of any answer. Experts in numerical analysis consider this to be a design flaw. Perhaps this observation lends weight to the FlySpeck project (a 20-year program to computer-certify the Hales–Ferguson proof).

A fascinating and detailed history of the Kepler sphere-packing problem, and related mathematical questions, appears in [ASW]. Ferguson and Hales have been awarded the 2007 Robbins Prize by the American Mathematical Society, and that seems to be a fitting reward for their achievements.

Exercise: Calculate the density of pennies for the rectilinear packing in the plane. Calculate the density for the hexagonal packing.

---

[4]This is also new territory for the venerable *Annals of Mathematics*. The *Annals* traditionally does *not* publish outlines.

[5]At a recent meeting, held in England, to discuss the changing nature of mathematical proof, MacPherson commented on the handling of the Hales paper by the *Annals*. He claimed that the *Annals* has a new policy of accepting computer-generated and computer-assisted proofs. There is no public record of such a decision. He further asserted that the usual refereeing process of the *Annals* had "broken down" in the case of the solution of the Kepler sphere-packing problem. The evidence, as described here, suggests that there was more at play than such a simple explanation would suggest.

8.6   The Platonic Solids

The classical term "Platonic solid" is used to denote a polyhedron in three-dimensional space with the property that (i) all faces are regular polygons and are congruent and (ii) the same number of faces meet at each vertex. An example is the *cube*. The cube has six faces that are squares. Three faces meet at each vertex.

Find all the Platonic solids.

What is amazing is that there are a total of just five Platonic solids, and we can find them all. To do so, we exploit the formula of Euler that we studied in Section 8.4:

$$V - E + F = 2.$$

Recall that V is the number of vertices, E the number of edges, and F the number of faces. We think of the edges and vertices of the Platonic solid as forming an admissible graph on the surface of the Platonic solid (which, topologically, is the same as a sphere).

Because we are describing a *regular polyhedron*, there are certain relationships among V, E, and F. We let m be the number of edges to a face, and let k be the number of edges that meet at any vertex.

First of all, since each face has m edges, we might think that $m \cdot F$ describes the total number of edges. However this is not quite right because each edge bounds two faces (one on each side of the edge) so we have counted each edge twice. Thus

$$E = \frac{m \cdot F}{2}. \tag{8.6.1}$$

Second, each face has m vertices (since it has m edges). But the quantity $m \cdot F$ counts each vertex k times, since each vertex is the meeting place of k edges, hence of k faces. Thus

$$V = \frac{m \cdot F}{k}. \tag{8.6.2}$$

Substituting (8.6.1) and (8.6.2) into Euler's formula yields

$$\frac{m \cdot F}{k} - \frac{m \cdot F}{2} + F = 2.$$

Multiplying through by $2k$ and factoring on the left side gives

$$F \cdot (2m + 2k - mk) = 4k. \tag{8.6.3}$$

It turns out that (8.6.3) is a very rich formula that tells us everything that we want to know. Now we proceed by numbered steps:

(1) It cannot be that both $m \geq 4$ and $k \geq 4$. If both inequalities held and if $m \geq k$ then

$$mk \geq 4m$$

and

$$mk \geq 4k.$$

Multiplying each inequality by $1/2$ and adding gives

$$mk \geq 2m + 2k.$$

This makes the left side of (8.6.3) less than or equal to zero, which is impossible (since the right side is positive). A similar contradiction would result if $k \geq m$. Thus, in what follows, we may assume that *either $m < 4$ or $k < 4$.*

(2) We cannot have either $k > 5$ or $m > 5$. First consider the case $k > 5$. Then, by the preceding step, $m \leq 3$. But $m = 1$ or $m = 2$ make no sense (since a polygonal face cannot have just one side or two sides). Thus, if $k > 5$, then $m = 3$. Putting this information into the left side of (8.6.3) gives

$$F \cdot (2 \cdot 3 + 2k - 3k) = 4k$$

or

$$F \cdot (6 - k) = 4k.$$

But $k > 5$ makes the left side less than or equal to zero, which is again impossible.

A similar argument, which we omit, shows that $m > 5$ cannot be allowed.

(3) We have learned so far that both m and k must be less than or equal to 5, and they cannot both be greater than or equal to 4. Thus there are only finitely many cases to consider:

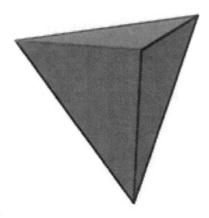

Figure 8.20: The tetrahedron.

$$m = 3 \qquad k = 3, 4, 5$$
$$m = 4 \qquad k = 3$$
$$m = 5 \qquad k = 3$$

Note that $m = 1, 2$ makes no sense geometrically, and neither do $k = 1, 2$. This limits our choices.

(**4**) We see that there are just five cases to consider, and each gives rise to a platonic solid:

(**a**) If $m = 3, k = 3$, then equation (8.6.3) yields that $F = 4$. This data corresponds to a polyhedron with four faces, each a triangle. And three triangles meet at each vertex. This is the *tetrahedron* (Figure 8.20).

(**b**) If $m = 3, k = 4$, then equation (8.6.3) yields that $F = 8$. This data corresponds to a polyhedron with eight faces, each a triangle. And four triangles meet at each vertex. This is the *octahedron* (Figure 8.21).

(**c**) If $m = 3, k = 5$, then equation (8.6.3) yields that $F = 20$. This data corresponds to a polyhedron with twenty faces, each a triangle. And five triangles meet at each vertex. This is the *icosahedron* (Figure 8.22).

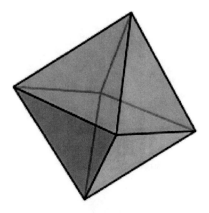

Figure 8.21: The octahedron.

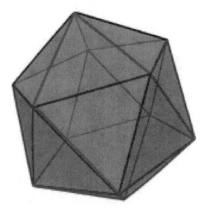

Figure 8.22: The icosahedron.

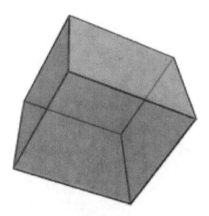

Figure 8.23: The cube.

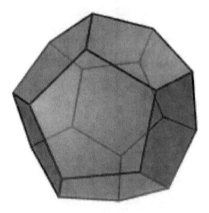

Figure 8.24: The dodecahedron.

(d) If $m = 4, k = 3$, then equation (8.6.3) yields that $F = 6$. This data corresponds to a polyhedron with six faces, each a square. And four squares meet at each vertex. This is the *cube* (Figure 8.23).

(e) If $m = 5, k = 3$, then equation (8.6.3) yields that $F = 12$. This data corresponds to a polyhedron with twelve faces, each a pentagon. And three pentagons meet at each vertex. This is the *dodecahedron* (Figure 8.24).

This completes our description of the five Platonic solids.

The situation for Platonic solids in higher dimensions is more complicated (and there are not as many of them). We cannot discuss it here.

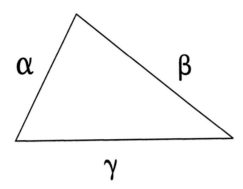

Figure 8.25: A triangle with sides labeled.

Exercise: In dimension 4 there are 6 Platonic solids while in dimension 5 there are 3 Platonic solids. Explore these situations and see whether you can describe the new Platonic solids.

8.7 Heron's Problem

Consider the problem of expressing the area of a triangle with a formula using only the lengths of the triangle's three sides.

We write the side lengths as α, β, γ as shown in Figure 8.25. We place the triangle on a coordinate system, as shown in Figure 8.26. Thus $\alpha = \sqrt{a^2 + c^2}$, $\beta = \sqrt{b^2 + c^2}$, and $\gamma = b - a$. Our plan is to express the area in terms of a, b, c, and then translate that to an expression involving only α, β, γ.

This problem is traditionally known as Heron's problem.

Obviously the area of the triangle is

$$
\begin{aligned}
A = \text{Area} &= \frac{1}{2} \cdot \text{base} \cdot \text{height} \\
&= \frac{1}{2} \cdot (b - a) \cdot c.
\end{aligned}
$$

Now we know that $b - a = \gamma$, so that our formula for area becomes

$$
A = \frac{1}{2} \cdot \gamma \cdot c.
$$

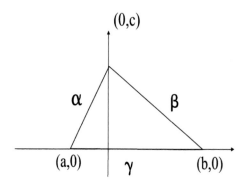

Figure 8.26: The triangle displayed in a coordinate system.

Our problem will be solved if we can express c in terms of α, β, γ.

In effect, we are simultaneously solving the equations

$$\begin{aligned}
\alpha^2 &= a^2 + c^2 \\
\beta^2 &= b^2 + c^2 \\
\gamma &= b - a
\end{aligned}$$

for the variable c in terms of α, β, γ. Of course these are not linear equations, so the problem does not bow to any standard technique. We will resort to fiddling (a technique that is *not* to be underestimated).

Looking at the figure, we see that the roles of α and β in the problem are symmetric. Thus we would expect our final expression for c to be symmetric in α and β. Thus it is natural to consider the expression $\alpha^2 + \beta^2$. We subtract γ^2 from it in order to cancel some terms. Thus

$$\alpha^2 + \beta^2 - \gamma^2 = 2c^2 + 2ab. \tag{8.7.1}$$

Another symmetric expression in α and β is $\alpha \cdot \beta$. However this will involve some nasty square roots so instead we consider $\alpha^2 \cdot \beta^2$. Thus

$$\alpha^2 \cdot \beta^2 = (a^2 + c^2) \cdot (b^2 + c^2) = a^2 b^2 + a^2 c^2 + c^2 b^2 + c^4. \tag{8.7.2}$$

Now if we expect to combine (8.7.1) and (8.7.2), then we have a mismatch. For the terms in (8.7.1) are of second order and those in (8.7.2) are of fourth order. Thus we consider

$$(\alpha^2 + \beta^2 - \gamma^2)^2 = (2c^2 + 2ab)^2 = 4c^4 + 4a^2 b^2 + 8abc^2. \tag{8.7.3}$$

Certainly there are a number of terms in (8.7.3) that also appear in (8.7.2). In order to obtain useful cancellations, we thus finally calculate

$$4[\alpha^2 \cdot \beta^2] - (\alpha^2 + \beta^2 - \gamma^2)^2$$

$$
\begin{aligned}
&= 4[a^2b^2 + a^2c^2 + c^2b^2 + c^4] \\
&\quad - [4c^4 + 4a^2b^2 + 8abc^2] \\
&= 4a^2c^2 + 4b^2c^2 - 8abc^2 \\
&= 4c^2[b-a]^2 \\
&= 4c^2\gamma^2.
\end{aligned}
$$

In summary,

$$c = \frac{\sqrt{4[\alpha^2 \cdot \beta^2] - (\alpha^2 + \beta^2 - \gamma^2)^2}}{2\gamma}.$$

Notice that we have succeeded in expressing c in terms of α, β, γ and nothing else.

Finally, the area of the triangle is

$$A = \frac{1}{2}\gamma \cdot c = \frac{1}{4}\sqrt{2\alpha^2\gamma^2 + 2\beta^2\gamma^2 + 2\alpha^2\beta^2 - \alpha^4 - \beta^4 - \gamma^4}. \qquad \square$$

Exercise: Notice that the formula that we have derived is symmetric in α, β, and γ. That is to say, if we permute the roles of α, β, γ, then the formula is unchanged. Think carefully about why this must be the case. Some books of tables (see [CRC]) give the area of the triangle in its more classical form:

$$A = \sqrt{s(s-\alpha)(s-\beta)(s-\gamma)},$$

where $s = (\alpha + \beta + \gamma)/2$, or half the perimeter of the triangle. Derive this classical form from the one that we have in the text.

8.8   A Little Geometric Reasoning

What is the greatest number of regions into which three straight lines (of infinite extent) can divide the plane?

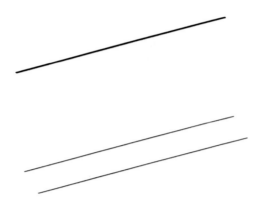

Figure 8.27: Two lines that coincide and two parallel lines.

We begin with the simpler question: "What is the greatest number of regions into which one line can divide the plane?" Of course there is nothing to discuss, for one line will always separate the plane into two regions.

Next we look at two lines. Refer to Figure 8.27 (top). If the two lines coincide, then the plane is still divided into just two regions. If, instead, the two lines are distinct but parallel (Figure 8.27, bottom) then the plane is separated into three separate regions.

We think of the two cases just described as degenerate or atypical for the following reason: If you drop two straws onto a floor, then the probability that they will land on top of each other, or land in a configuration so that they are parallel, is zero. Rather, with probability 1, the straws will land so that they are skew (or non-parallel). We refer to this last situation as "general position" for the two straws.

Now suppose that our two lines are in general position. This situation is illustrated in Figure 8.28. Then the plane is separated into four regions.

Finally we pass to three lines. If all three lines coincide, then we are in the situation for one line. If two of the lines coincide, then we are in the situation for two lines. So suppose that the three lines are distinct.

If the three lines are parallel, then the plane is separated into four regions (Figure 8.29). If two are parallel, and the third is skew to them, then the plane is separated into six regions (Figure 8.30). Now suppose that no two of the three lines are parallel.

If the three lines pass through a single point, then (Figure 8.31) the plane

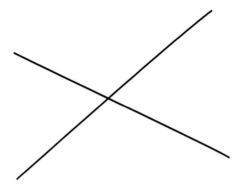

Figure 8.28: Two lines in general position.

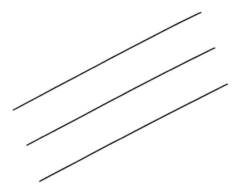

Figure 8.29: Three parallel lines.

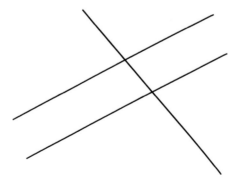

Figure 8.30: Two parallel lines and one skew.

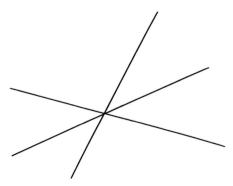

Figure 8.31: Three lines through a single point.

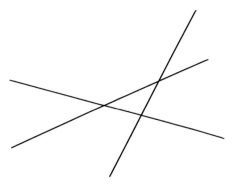

Figure 8.32: Three lines in general position.

is separated into six regions. If the three lines do not pass through a single point, and no two of them are parallel (this is the general position situation, that is, the one that occurs with probability 1—see Figure 8.32), then the plane is separated into seven regions. Thus seven is the maximal number of regions into which three lines can divide the plane.

Exercise: The maximal number of regions into which 3 planes can divide 3-dimensional space is 8. The maximal number of regions into which 4 planes can divide 3-dimensional space is 15. Explore these two assertions.

Chapter 9

Practical Matters

9.1 Strangers on a Plane

At a cocktail party at my university there was a spirited conversation about whether academics are more eccentric and depressive than other highly motivated professionals. At some point, a psychologist who had studied the matter extensively joined the conversation and declared that, no, academics were definitely *not* more eccentric and depressive than other professionals. The only exceptions, he noted, were mathematicians and oboe players.

It seems that the property that mathematicians and oboe players have in common is that they both do something very difficult that nobody else appreciates. Certainly anyone who is a professional mathematician has become accustomed to meeting people who respond to his/her being a mathematician by saying, "Oh, mathematics was my worst subject." Or perhaps, "I was really good at math until we got to that stuff with the letters—like algebra. Then I was lost." Distinguished mathematician Mary Ellen Rudin has observed that this would be like saying to an English professor, "Oh, I was really good at English as long as we stuck to *A, B, C*. But when we got to *D, E, F* then I was lost."

Recent developments suggest that perhaps the general public is becoming more copacetic with mathematics. The enthusiastic response to the solution of Fermat's Last Theorem, the response to the proof of the Poincaré conjecture, the popularity of the TV show *Numb3rs* all suggest at least a public awareness of mathematics. All parts of life are now becoming mathematized. Mathematicians are enjoying a little of the respect and admiration that we all believe we deserve.

But there are some notable exceptions. One Thursday evening, a 40-year-old man—with dark, curly hair, olive skin, and an exotic foreign accent—boarded a plane. It was a regional jet making a short hop from Philadelphia to Syracuse.

The curly-haired man tried to keep to himself, intently if inscrutably scribbling on a notepad he had brought aboard. His seatmate, a blond-haired, relatively young woman, eyed him skeptically. He was wearing navy Diesel jeans and a red Lacoste sweater, but something about him did not seem right to her.

She decided to try out some small talk. "Is Syracuse home?" She asked. "No," he replied curtly.

He deflected further questions. He appeared laser-focused—perhaps too laser-focused—on his scribblings on his notepad.

Rebuffed, the woman began reading her book. Shortly after boarding was complete, she flagged down a flight attendant and handed that crew-member a note.

Then the passengers waited, and waited, and waited for the flight to take off. After they had sat on the tarmac for about half an hour, the flight attendant approached the female passenger again and asked if she now felt okay to fly, or if she was "too sick." "I am OK to fly," the woman said. But American Airlines Flight 3950 remained grounded.

Then the plane turned around and headed back to the gate. The woman was soon escorted off the plane. On the intercom a crew member announced that there was paperwork to fill out, or fuel to refill, or some other mysterious circumstance.

The wait continued.

Finally the pilot came by, and approached the darkly complected foreign man. He was now escorted off the plane and taken to meet an agent. "What do you know about your seatmate?" the agent asked the man. "Well, she acted a bit funny," he replied.

Then the secret was revealed. The woman, his seatmate, had seen the cryptic notes the man had written, scrawled in a script she did not recognize. Maybe it was code, or some foreign lettering, possibly the details of a plot to destroy the dozens of innocent lives aboard American Airlines Flight 3950. She felt it her duty to alert the authorities. The curly-haired man was, the agent informed him politely, suspected of terrorism.

The curly-haired man laughed. He laughed because those scribbles were not Arabic, or another foreign language, or even some special secret terrorist code. They were math. Yes, math. A differential equation, to be exact.

Had the crew or security members perhaps quickly `Googled` this good-natured, bespectacled passenger before waylaying everyone for several hours, they might have learned that he—Guido Menzio—is a young but distinguished Ivy League economist. And that he is best known for his relatively technical work on search theory, which helped earn him a tenured Associate Professorship at the University of Pennsylvania as well as stints at Princeton and the Hoover Institution of Stanford.

Guido Menzio, an economics professor at the University of Pennsylvania. Not an Arab. They might even have discovered that last year he was awarded the prestigious Carlo Alberto Medal, given to the best Italian economist under 40.

Menzio had been on the first leg of a connecting flight to Ontario, where he would give a talk at Queens University on a working paper he co-authored about menu costs and price dispersion. His nosy neighbor had spied him trying to work out some properties of the model of price-setting he was about to present. Perhaps she could not differentiate between differential equations and Arabic.

I have been a mathematician for 45 years, and I have frequently done mathematics on airplanes. But I am happy to say that I have never been harassed as a result. I once sat next to a retired professional baseball player and we had a nice chat about the catalog of expensive cars that he was perusing. But when I brought out my math to while away the remaining time on the flight he acted offended and moved away to another seat. That is my one brush with fame.

Exercise: Pure mathematicians are often accused of working on problems that have no practical significance. G. H. Hardy, in his famous book *A*

Mathematician's Apology, [HAR] crowed that he had never done anything practical. But, these days, new applications of pure mathematics are being discovered every day. The design of the memory chip in your computer is based on abstract mathematical ideas. So is the engineering behind music CDs. Guido Menzio was looking at abstract mathematics but applying it to economics. Go on the Internet and find ten applications of abstract mathematics to practical situations.

9.2   You've Got My Vote

Elections are much more complicated than you may think. A famous theorem of Kenneth Arrow says that any election system may be manipulated (that is, a surprising and perhaps undesirable outcome may be forced by rigged voting). You may not have known that there is more than one election system. But there is. We will learn about some of them in this section.

When there are just two candidates, and every voter votes for just one of them (i.e., there are no write-ins and no abstentions), then life is simple: Whichever candidate gets the most votes wins. End of discussion.

But when there are three or more candidates life can get complicated. Just as an instance, in 1824 John Quincy Adams was elected president of the United States, even though he did not get the greatest number of popular votes nor the greatest number of votes in the Electoral College. How could this be?

In point of fact, Adams received 31% of the popular vote while Andrew Jackson received 41%. Also Adams received 84 electoral votes while Andrew Jackson received 99 electoral votes. Recall that, in the United States, the president is *not* elected by the popular vote. The popular vote determines who certain "electors" will be, and the electors select the president. In the 1824 election, William H. Campbell received 41 electoral votes and Henry Clay received 37 electoral votes. So Jackson had a plurality of electoral votes, but definitely not a majority.

According to the Twelfth Amendment to the Constitution of the United States, if no candidate receives a majority of the electoral votes then the election is thrown to the House of Representatives. Under the influence of politicking by Henry Clay, the House of Representatives thus elected John

Quincy Adams to be president. You can imagine that Andrew Jackson and his supporters were very unhappy about this outcome.

An even more bizarre situation took place in the 2000 presidential election. The candidates at that time were George W. Bush and Al Gore. In fact, Gore received at least 540,000 more popular votes than Bush. But, following a Supreme Court order, Bush was ultimately awarded 5 more electoral votes than Bush. So Bush won the election. Some have argued that Ralph Nader's involvement in the election skewed the vote and stole the victory from Gore.

This last point is worth exploring a bit. Bush beat Gore in New Hampshire by less than 8,000 popular votes. But Nader received about 22,000 votes in New Hampshire. In Florida, Bush beat Gore by just a few hundred popular votes. But Nader received about 97,000 votes. If the Nader supporters had voted for one of the two major candidates, then the race could easily have gone the other way. Of course Nader never expected to be elected president; he just wanted to make a statement. And he certainly did.

In fact, the number of electors that each state gets is equal to the number of senators from that state plus the number of congressmen from that state. Every state has two senators and at least one congressman. So every state has at least three electors. If the presidency were decided by popular vote, then it is likely that candidates would ignore the small-population states. But three electors really counts. So candidates campaign in all the states.

It is interesting to calculate the number of residents per elector in the various states. Some of the small-population states, such as Vermont, have about 200,000 residents per elector (according to the 2000 census). For South Dakota the number is about 250,000 residents per elector, and for North Dakota it was about 214,000. Contrast this with California, where there were 616,924 residents per elector. For New York it was 613,064 residents per elector. We may conclude from these numbers that residents of the small-population states have considerably more influence over the presidential election than do residents of the large-population states. It has been argued that the small states favor the electoral college system since they derive empowerment from that system.

Let us now examine several commonly used voting systems and compare and contrast them for their effectiveness. One of the more interesting examples below is the Hare system, which is an instance of a system of proportional representation. In fact the Hare system is used in Australia, Malta, Ireland,

and Cambridge, Massachusetts. It is used in several professional societies in the United States (including the American Mathematical Society). We begin with the plurality system.

9.2.1   The Plurality System

This is perhaps the most commonly used voting method. In it you have several (perhaps more than two) candidates, and the candidate who gets the most votes wins. Period.

For example, suppose that a certain election has candidates Smith, Jones, and Brown. Imagine that there are 10,000 voters in this election, and their votes are distributed as follows:

Name	No. of Votes
Brown	3750
Jones	4010
Smith	2240

By the plurality method, Jones wins the election—even though Jones does not have a majority of the votes. Put in other words, the majority of the voters *do not* prefer Jones. But she wins anyway. Many people would say that this voting system is unfair.

9.2.2   The Hare System

This is sometimes called the system of "single transferable vote" (STV). Although the Hare system has a number of weaknesses, it also has several strengths.

We shall describe how the Hare system works on the fly, in the context of a particular example below.

Imagine an election with four candidates: Smith, Jones, Brown, and

Xbibtl. There are 17 voters, and they rank the candidates as follows

No. of Voters	1st Choice	2nd Choice	3rd Choice	4th Choice
6	Xbibtl	Brown	Jones	Smith
6	Xbibtl	Jones	Smith	Brown
5	Xbibtl	Smith	Brown	Jones

We assume that there are two candidates to be elected.

A focal point in these discussions is the *Droop quota*. If n is the number of voters and m is the number of positions to be filled, then the Droop quota q is

$$q \equiv \frac{n}{m+1} + 1.$$

This quotient may not be a whole number, so in practice we use the greatest integer in this quotient, or

$$\tilde{q} \equiv \left[\frac{n}{m+1} + 1 \right].$$

What is the significance of this number? It is the smallest integer $\tilde{q}$ so that not more than m of the candidates can get $\tilde{q}$ votes.

For the set problem described in the third paragraph,

$$\tilde{q} = \left[\frac{17}{3} + 1 \right] = \left[\frac{20}{3} \right] = 6.$$

Indeed, two candidates could get 6 votes ($6 + 6 = 12$) but three could not (because $6 + 6 + 6 = 18 > 17$).

Now here is an analysis of this election using the Hare system:

First Round: Candidate Xbibtl receives 17 out of 17 first-place votes. So she is elected. But remember that we are looking to elect two people.

Second Round: Candidate Xbibtl has a surplus of 11 votes (over the requisite $q = 6$). So these must be distributed to the other three candidates in the proportions 6/17, 6/17, 5/17 (notice that these add up to 1). So two of the candidates receive 3.9 votes and one receives 3.2 votes.

The candidate with the fewest transfer votes is automatically eliminated in the Hare system. So Smith is out. We see in the third line of the table that the next choice after Smith is Brown. So Smith's 3.2 votes go to Brown. And $3.9 + 3.2 = 7.1$ exceeds the cutoff of 6. Thus Brown gets the second slot. So our two winners are Xbibtl and Brown.

Now let us modify the problem and see how it changes. Namely, let us suppose that 2 of the 6 voters who ranked the candidates

Jones, Smith, Brown

still indicate Xbibtl as their first choice but do not specify a second or third or fourth choice. Now what happens?

First Round: Candidate Xbibtl receives 17 out of 17 first place votes. So she is elected. But remember that we are looking to elect two people.

Second Round: Candidate Xbibtl has a surplus of 11 votes (over the requisite $q = 6$). So these must be distributed to the other three candidates in the proportions 6/17, 4/17, 2/17, 5/17 (notice that these add up to 1). There are now four numbers because the second group of 6 voters has been broken up into 4 and 2. Thus

$\frac{6}{17} \cdot 11 = 3.9$ votes go to Brown

$\frac{4}{17} \cdot 11 = 2.6$ votes go to Jones

$\frac{5}{17} \cdot 11 = 3.2$ votes go to Smith

The candidate with the fewest transfer votes—Jones—is automatically eliminated in the Hare system. The supporters of Jones transfer their votes to the next highest choice, which is Smith. So Smith now has $3.2 + 2.6 = 5.8$ votes. But $5.8 < 6$. So this is inconclusive.

On the other hand, $3.9 < 5.8$. So Smith beats Brown and Smith is now the second elected candidate. The winners are Xbibtl and Smith.

One can see here that there are advantages to not playing the STV game and just voting for a single candidate (not specifying second choice, third choice, etc.).

We next give an example with just one candidate to be selected. We will see new flaws in the STV system.

We shall assume that there are four candidates: Xbibtl, Brown, Jones, and Smith. There are 21 voters, and the votes are distributed as follows:

No. of Voters	1st Choice	2nd Choice	3rd Choice	4th Choice
7	Brown	Jones	Smith	Xbibtl
6	Jones	Brown	Smith	Xbibtl
5	Smith	Jones	Brown	Xbibtl
3	Xbibtl	Smith	Jones	Brown

We calculate that

$$\tilde{q} = \left[\frac{21}{1+1} + 1 \right] = \left[\frac{23}{2} \right] = 11 .$$

Because no candidate has a simple majority of $\tilde{q} = 11$ first-place votes, the lowest first-choice candidate, Xbibtl, is eliminated on the first round. Looking at the fourth line of the table, we see that the second-place vote goes to Smith. This gives Smith $5 + 3 = 8$ votes. Because none of the remaining candidates has a majority at this point, we see that Jones, with the new lowest total of 6 voters, is now eliminated. And Jones's second-place votes go to Brown. Thus Brown is elected with a total of $7 + 6 = 13$ votes.

If instead 3 of the voters in the fourth row of the table indicate only Xbibtl as their first choice, then Xbibtl is still eliminated on the first round. But since the voters in this row did not indicate a second choice, no votes are transferred. Now Smith is the new lowest candidate, with 5 votes. So Smith is eliminated, and his/her votes are then transferred to Jones. And Jones is then elected with $6 + 5 = 11$ votes. Because the voters in row 4 prefer Jones to Brown, it is in their interest not to rank candidates below Xbibtl in order to induce a better outcome for themselves.

We conclude this subsection with an example, for which we thank Don Saari, in which the Hare system gives a startling and counterintuitive result.

Imagine an election in which there are 99 voters and seven candidates called A, B, C, D, E, F, G. We seek to elect 3 people. Assume that 33 of the voters favor the candidates in this order:

- A

- B
- E
- F
- G
- C
- D

Also 33 of the voters favor the candidates in this order:

- C
- B
- F
- E
- G
- D
- A

And finally, 33 of the voters favor the candidates in this order:

- D
- B
- E
- F
- G
- A
- C

We see that candidate B is very popular. But she will not get a single vote. In fact A, C, D win without even any vote transfers at stage two. Note that in fact D and A are ranked *last* by two of the voters!

9.2.3 The Borda Count

The Borda system was proposed over 200 years ago by Jean-Charles de Borda (1781). Generally speaking, it is not used to elect candidates to public office. But it is used by many private organizations.

In the Borda system, points are assigned to each candidate so that the lowest-ranking candidate of each voter receives 0 points, the next-lowest candidate receives 1 point, and so on up to the highest-ranking candidate who receives $m - 1$ points if there are m candidates. Points for each candidate are summed across all voters, and the candidate with the most points wins.

In the analysis that follows, we shall refer to the *Condorcet candidate* as that candidate who would beat each of the other rivals in a one-on-one race.

First consider a race in which three voters prefer the candidates in the order Brown, Jones, Smith. But two voters prefer the candidates in the order Jones, Smith, Brown. Clearly Brown is the Condorcet candidate (because three of the voters rank him/her first). But the Borda score of Brown is $3 \cdot 2 + 2 \cdot 0 = 6$, the Borda score of Jones is $3 \cdot 1 + 2 \cdot 2 = 7$, and the Borda score of Smith is $3 \cdot 0 + 2 \cdot 1 = 2$. So Jones is the Borda winner. Note that the Borda winner is different from the Condorcet winner.

Now review the example above (in the Hare subsection) in which we had four candidates and just one winner. In that example, the Borda score of Brown is $7 \cdot 3 + 6 \cdot 2 + 5 \cdot 1 + 3 \cdot 0 = 38$, the Borda score of Jones is $6 \cdot 3 + 12 \cdot 2 = 42$, the Borda score of Smith is $5 \cdot 3 + 3 \cdot 2 + 13 \cdot 1 = 34$, and the Borda score of Xbibtl is $3 \cdot 3 + 18 \cdot 0 = 9$. So we see that Jones is the clear Borda winner. Even though Brown is the Condorcet winner. And the reason is that Jones has the highest average position in the ranking:

Name	Average Ranking	
Brown	$\frac{7 \cdot 1 + 6 \cdot 2 + 5 \cdot 3 + 3 \cdot 4}{21}$	≈ 2.19
Jones	$\frac{6 \cdot 1 + 12 \cdot 2 + 3 \cdot 3}{21}$	≈ 1.86
Smith	$\frac{5 \cdot 1 + 3 \cdot 2 + 12 \cdot 3}{21}$	≈ 2.24
Xbibtl	$\frac{3 \cdot 1 + 0 + 0 + 18 \cdot 4}{21}$	≈ 3.57

We see that Jones has the average ranking that is nearest to 1, confirming our Borda calculation. Thus the Borda ranking is an indication that the winning candidate is broadly acceptable to voters, unlike the Condorcet candidate.

The Borda count can be manipulated in some surprising ways. We illustrate this observation with an example.

Imagine that we have seven voters and the usual three candidates (Brown, Jones, and Smith). Let us lay out their voting preferences and their Borda counts:

No. of Voters	1st Choice	2nd Choice	3rd Choice
3	Smith	Jones	Brown
2	Brown	Smith	Jones
2	Jones	Brown	Smith

We see that the Borda counts are

Brown: $3 \cdot 0 + 2 \cdot 2 + 2 \cdot 1 = 6,$

Jones: $3 \cdot 1 + 2 \cdot 0 + 2 \cdot 2 = 7,$

Smith: $3 \cdot 2 + 2 \cdot 1 + 2 \cdot 0 = 8.$

Thus Smith wins by the Borda method.

But now suppose that we introduce a sure loser—namely Xbibtl—into the mix. Now the ranking is

No. of Voters	1st Choice	2nd Choice	3rd Choice	4th Choice
3	Smith	Jones	Brown	Xbibtl
2	Brown	Xbibtl	Smith	Jones
2	Jones	Brown	Xbibtl	Smith

Notice that we have made no change in the relative rankings of Brown, Jones, and Smith. But we have inserted Xbibtl into the ranking. Now the Borda scores are

Brown: $3 \cdot 1 + 2 \cdot 3 + 2 \cdot 2 = 13,$

Jones: $3 \cdot 2 + 2 \cdot 0 + 2 \cdot 3 = 12,$

Smith: $3 \cdot 3 + 2 \cdot 1 + 2 \cdot 0 = 11$.

So we have in fact reversed the order of the Borda scores! Notice that the Borda score of Xbibtl is $3 \cdot 0 + 2 \cdot 2 + 2 \cdot 1 = 6$, which is why we say that Xbibtl is a sure loser, or perhaps "irrelevant."

9.2.4 Cumulative Voting

The cumulative voting strategy has been used for elections to the Illinois General Assembly and by the cities of Alamogordo, New Mexico, and Peoria, Illinois. It has been used for elections to some corporate boards of directors. It is perceived to satisfy court requirements of minority representation in municipal elections.

The way that it works is this. Each voter is given a certain number of votes (usually more than one). And he/she can cast them in any possible manner—all votes for one candidate, or split among the candidates. Thus a voter can express intensity of preference. This is a system of proportional representation.

Cumulative voting is best understood by way of an example. Suppose that we have 300 voters, and they are to elect 6 members to a governing board. Each voter has 6 votes to cast. There is a minority group of 100 among the voters that favors a particular voting pattern, and a majority group of 200 voters that favors an opposing voting pattern.

If each of the 100 voters in the minority casts 3 votes each for two particular candidates, then each of those two candidates will receive 300 votes. If, in opposition, the minority of 200 supports 4 different candidates, then they have $200 \cdot 6 = 1200$ votes to distribute among the 4 candidates. So the best that they can do is give 300 votes to each candidate. So they will tie the two minority candidates.

If instead the two-thirds majority decides to distribute its votes equally among 5 candidates, then each candidate will get $1200/5 = 240$ votes, and will lose to the 300 votes accrued by the two minority candidates. If instead the minority decides to support 3 candidates, then each will receive 200 votes and they will lose.

Clearly the optimal strategy for each group is to distribute its aggregate votes as equally as possible among a certain number of candidates. Any deviation from this strategy, say by putting up a full slate of candidates and not instructing supporters to vote for only some of the slate, gives the other side the opportunity to grab more than its proportional share of seats.

9.2.5   Approval Voting

Some experts in the mathematical theory of voting think that approval voting is the best system. Many professional societies use this system. Some political parties in some states use approval voting for internal elections. Approval voting has been used for some elections in the Soviet Union. The major political parties in the United States have not been persuaded to use it on the national level, even though it would certainly work in their favor in some circumstances.

The system is simplicity itself. Each voter is allowed to cast as many votes as he/she pleases. But just (at most) one vote for each candidate. Some advantages of approval voting are the following:

- **Voters are given more flexible options.** If the voter has no strong preference for one candidate, he/she can express this by voting for all candidates found to be acceptable.

- **Increased voter turnout.** Since voters are better able to express their preferences, they are more likely to go to the polls. The election system allows them to be more honest.

- **The system helps to elect the strongest candidate.** With this system, the candidate with the greatest overall support wins.

- **The system gives minority candidates their proper recognition.** Voters would not be attracted away from a minority candidate because there was another candidate who, though less appealing, was considered to be a stronger contender.

- **The system is practicable.** The system is easy to implement on voting machines, and easy for voters to understand.

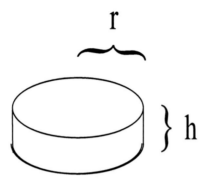

Figure 9.1: A tablet as a right circular cylinder.

9.2.6   Conclusions

No voting system is perfect. But different voting systems have different strengths. By understanding the different features of the voting systems, we can choose one that most nearly meets our personal and professional goals.

Exercise: Kenneth Arrow's theorem says that any voting system can be manipulated. Pick one of the voting systems described in this section and explain how it can be manipulated.

9.3   Take Your Pill

Traditionally all pills (commonly known in the business as "tablets") were in the shape of a right circular cylinder. See Figure 9.1.

Those who manufacture tablets of course want to know the volume and the surface area of any given pill. They need to know the volume so that they can be sure that the right amount of medicine is contained in the pill. They need to know the surface area so that they can have an idea of the absorption rate of the medicine.

Fortunately, it is easy to calculate these quantities for a right circular cylinder. The area of a circular face of the tablet (see Figure 9.1) is πr^2 and hence the volume is $V = h \cdot \pi \cdot r^2$. The surface area of the tablet is twice the area of a face plus $2\pi rh$ for the area of the side. So $A = 2\pi r^2 + 2\pi rh$.

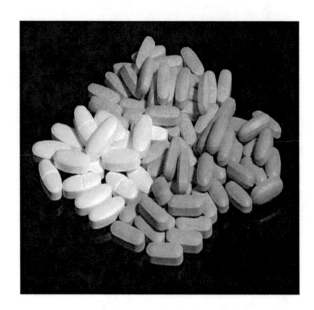

Figure 9.2: Some caplets.

In the last fifty years, fashions have changed in the pill industry. Now every manufacturer wants the tablets that they produce to have a character-istic and recognizable shape. Typically the shape is a *caplet.* See Figures 9.2 and 9.3.

Now the faces of the tablet are curved in a nontrivial fashion, and the basic shape is no longer circular. It is a difficult and complicated problem to calculate the volume and surface area of such an object.

The biggest producer in the world of machine dies for stamping out tablets is located in the St. Louis, Missouri area (home of the author of this book). What they did for many years to address the volume area problem for caplets is this. A customer would order machine dies to produce a certain tablet in a certain shape. And the machine die maker would produce the dies in five different sizes and let the customer choose which one was best for them. In other words, nobody actually calculated the volume or surface area because they simply did not know how to do it.

They needed to address this problem in a more scientific fashion. So, once the microcomputer age hit, what they did was to purchase a very complicated and expensive CAD (Computer-Aided Design) system from the high-tech company McDonnell-Douglas. The program lived on many dozens of discs

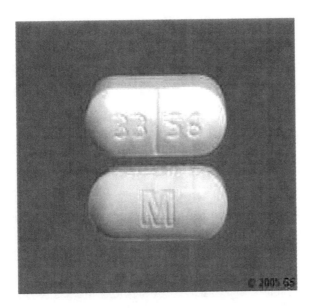

Figure 9.3: Some caplets.

(this was really the old days of computing) and the documentation ran to hundreds of volumes. A specially trained operator would render an image of the caplet on the screen, and then use a probabilistic method to bounce a particle around inside the caplet and thereby "calculate" the volume. The operator was unable to calculate the surface area.

The solution of the problem that we have just described was not considered to be very satisfactory. Because the CEO of the company was frequently on the phone with his customer in Germany or Singapore talking about the design of tablets, he wanted to be able to calculate the volume and surface area of a tablet in real time, during the phone conversation. The process with the CAD took several hours—sometimes days. This was simply not a workable system.

Enter the mathematician. The machine die company hired this author as a consultant to calculate the areas and the volumes. In fact, he could actually calculate some of them by hand, just using methods of calculus. So he could provide the machine die company with a formula for the volume (area was too hard). But for most caplets he could not do this. So he wrote a computer program to calculate the areas and the volumes. And he created a graphic of the caplet. The user could choose from among twelve different caplet designs. The software then prompted him for each dimension—shown clearly on the graphic. All the user had to do was enter the dimensions of the caplet, and the software would instantly spit out the area and the volume. It would also make recommendations about pressure to be applied in manufacturing the pill.

This creation revolutionized the pharmaceutical industry. No other machine die company in the world had any software like this. Suddenly the St. Louis company was on the cutting edge, and all the pill makers in the world were coming to them for the latest in pill manufacturing technology.

This is a very simple example of how our world today is becoming ever more mathematical. All different aspects of our life, from the carburetion system of the Volvo to the method for programming music CDs to the design of the hulls of America's Cup yachts are strongly influenced by mathematical thinking and mathematical ideas. A student studying mathematics today has a great panorama of career opportunities—from working in a genome lab to working for a high-tech company (like Microsoft or Texas Instruments) to working for a government research lab (like Los Alamos) to working for the National Security Agency. Or working at a college or university. This is an exciting time for mathematics.

Exercise: Design a pill shaped like a right circular cylinder that will have volume .1 cubic inches and surface area .25 square inches. Give the precise dimensions of such a pill.

9.4   Geometric Analysis and Facial Structure

Facial shape and its apotheosis, beauty, have fascinated man for thousands of years. Analysis of the face and, concomitantly, the quantification of beauty are subjects that have wanted for effective mathematical analysis. Today, these have become avenues of inquiry that are additionally driven by significant practical needs. National security issues make it essential that we have highly developed facial recognition technology for identifying and classifying terrorists and others who pose a threat to our well-being.

And problems of plastic surgery—in both its guises: **(i)** cosmetic and **(ii)** reconstructive, or the repair of damaged or deformed body parts—give rise to differential-geometric problems of analysis and design. Significantly in our digital age, facial shape seems to be one of the few areas that has thus far resisted all but the most limited quantification. The primary reason for this state of affairs is the essentially infinite dimensionality of the space of all possible faces.

The literature of facial recognition, analysis, and design is vast and highly diffuse. But some unifying themes are (i) much of the work, especially in the medical arena, is restricted to two-dimensional geometric considerations; (ii) most of the work is combinatorial, naively analytical, or computer driven, rather than differential geometric; (iii) most of the work does not address the fundamental issues of registration, isolation of key features, and metric comparison. The project that we describe below will address these missing design features. We base our approach on the fundamental observation that the face is, mathematically, a two-dimensional geometric surface, and there-fore subject (at least in principle) to differential geometric analysis. It can also be thought of as the graph of a function and hence is amenable to ana-lytic tools such as wavelet transforms and harmonic analysis decomposition.

We have a three-dimensional, white light scanner that operates much like an ordinary analogue camera. It will render the face as a finely triangulated surface; in fact, the vertices of the triangulation are recorded as a *.wrl file. See Figures 9.4, 9.5, and 9.6. Such data are grist for the mathematician's mill. This amounts to a detailed mathematical description of the facial surface—one that we can analyze.

9.4.1 Geometry and Facial Structure

We apply methods of differential geometry, Fourier and wavelet analysis, and mathematical real variable theory to analyze and quantify the shape of the face. See Figure 9.7, which exhibits regions of curvature using a shade coding. Figures 9.8, 9.9, and 9.10 also illustrate these ideas.

9.4.2 Conformal Mapping

Certain of our ideas have an interesting and inspiring history. The clas-sic work *On Growth and Form* by D'Arcy Thompson [THO]—now over 100 years old—posited that evolutionary development was linked to the mathe-

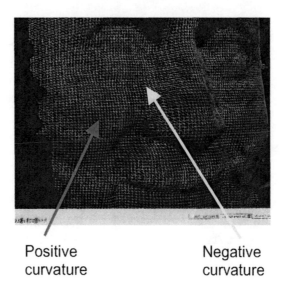

Positive   Negative
curvature   curvature

Figure 9.4: The *.wrl file from a facial scan.

matical idea of conformal mapping. These ideas have direct application to considerations of reconstructive surgery. One of our goals is to apply analytical ideas to develop planning techniques for plastic surgery procedures (such techniques are virtually nonexistent today). Certainly the comparison of the "goal" face with the existing face is an exercise in conformal mapping. Figures 9.11, 9.12, and 9.13 illustrate Thompson's seminal ideas.

9.4.3   Wavelets and Filters

The modern theory of wavelets—an outgrowth of the more classical idea of Fourier series—allows us to decompose any surface, in particular the face, into fundamental units. These units may be examined individually, or assembled to view substrata of the facial structure. Thus, in effect, we may construct filters that allow us to focus on various coarse and fine structures of the face. Our fundamental view is to enable the plastic surgeon to see things that he has never seen before. In the end, he/she is the skilled technician who will make the critical decisions and carry out the procedures. But we will provide him/her with powerful tools that will inform the process. Figures 9.15 and 9.16 illustrate the idea of extracting information at different "frequencies" from a facial structure.

Figure 9.5: More *.wrl files.

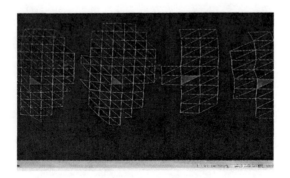

Figure 9.6: Additional *.wrl files.

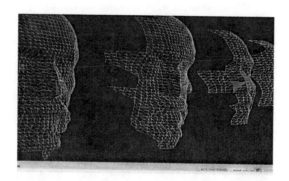

Figure 9.7: Comparison of goal face with existing face.

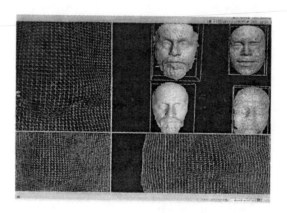

Figure 9.8: Applications of differential geometry.

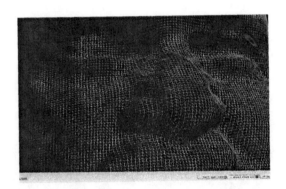

Figure 9.9: Comparison of four faces.

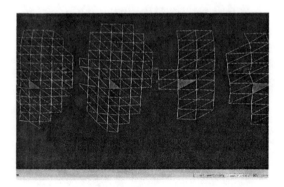

Figure 9.10: Analysis of curvature.

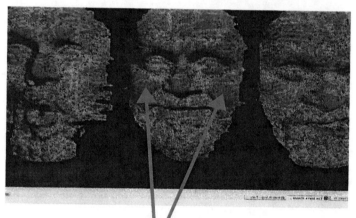

Smile causes cheeks
to become rounder
hence more positive
curvature.

Figure 9.11: Ideas of D'Arcy Thompson.

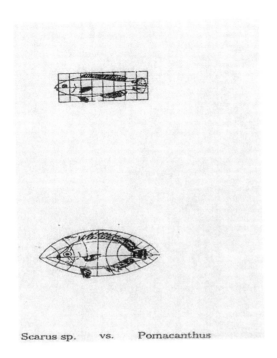

Scarus sp.   vs.   Pomacanthus

Figure 9.12: More from D'Arcy Thompson.

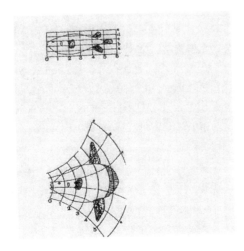

Figure 9.13: Comparison of different frequencies.

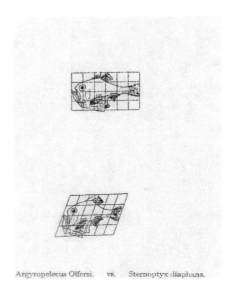

Argyropelecus Olfersi.   vs.   Sternoptyx diaphana.

Figure 9.14: A coarse frequency.

In focusing on coarse features of the face, the reconstructive surgeon can make global decisions about the changes that will be effected by the planned surgical procedure. By examining the fine features, the surgeon can perfect the procedural plan and insure that all the details of the face cohere and produce an attractive whole. Refer to Figures 9.14 and 9.15.

9.4.4   Summary Remarks

Traditionally the art of reconstructive surgery has been viewed as a form of craftsmanship. More like sculpture than science, reconstructive surgery consisted of fashioning a new face or other bodily part from an artistic vision of how things "ought to be."

But reconstructive surgery can be and should be an exacting analytical science. Techniques of differential geometry and mathematical analysis can be used to quantify and analyze the parameters that will affect a surgical procedure. Our effort is to turn these observations into a palpable science that will inform the surgeon as he designs and implements surgical procedures.

And it may be noted that these observations apply as well to other branches of medicine. Much of medical science is craft rather than analytical

Figure 9.15: Coarse features versus fine features.

discipline. The surgeon relies more on his/her experience than on objective data. He/she uses heuristic observation when he/she should be using hard evaluation of concrete information. We hope to change this paradigm so that the end results of surgical procedures can be relied on in the same way that the end results of business decisions or other objectifiable experiments can be assessed. We provide hard, objective, reliable analytical tools with reproducible and evaluable results and procedures. We hope to change the way that medical procedures are practiced, and to produce a climate in which medicine is less a matter of faith and more a matter of analysis (Figures 9.16, 9.17).

Our closing figure—Figure 9.18—represents the way that the physician will view the face in the future.

Exercise: What is the most popular plastic surgery procedure for men? What is the most popular plastic surgery procedure for women? Why are these different?

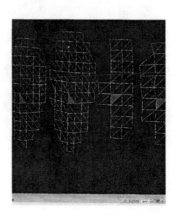

Figure 9.16: More coarse features.

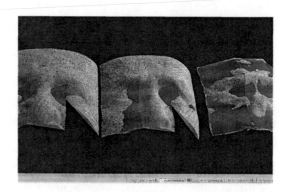

Figure 9.17: Coarse features versus fine features.

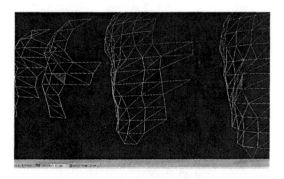

Figure 9.18: Coarse features.

9.5   Beware the Raven

This section touches on ideas from logic and from game theory. The mathematical theory of games dates back to work of O. Morgenstern and J. von Neumann in 1944 (see [MOV]). Of course logic has been studied at least since the time of Aristotle.

We begin by discussing the *raven paradox*, also known as *Hempel's paradox*. This logical conundrum arises from the question of what constitutes evidence for an assertion. Consider for example the assertion that all ravens are black. See Figure 9.19. What evidence might tend to support this claim?

First, note that the statement

$$\text{All ravens are black.} \tag{9.5.1}$$

is logically equivalent to the statement

If something is not black, then it is not a raven.

Thus, observing an object that is not black and not a raven may be thought to support the assertion (9.5.1). For example, observing a red Volksvagen

Figure 9.19: A black raven.

might be thought to logically validate the contention that all ravens are black. But this claim runs counter to our intuition. What does looking at a red Volksvagen have to do with, or tell us about, ravens?

It was the logician Carl Gustav Hempel who, in the 1940s, formulated this problem to illustrate a contradiction between inductive logic and intuition.

Nicod's criterion asserts that only observations of actual ravens should affect one's opinion as to whether all ravens are black. Observing additional instances of black ravens should support the claim, while observing white or colored ravens should contradict it. Observations of non-ravens should have no influence on the outcome, or the conclusion.

Now the set of all ravens is finite. By contrast, the set of all non-black things is either infinite or beyond human evaluation. In order to verify the statement "All ravens are black," we would have to look at all ravens. This is feasible. On the other hand, to verify the statement "All non-black things are non-ravens" (which of course is logically equivalent), it would be necessary to examine all non-black things. This is infeasible. Noticing a black raven can be thought of as a finite and nontrivial amount of confirmatory evidence for assertion (9.5.1). But observing a non-black non-raven is an infinitesimal amount of evidence.

I. J. Good [GOO] has argued that the observation of a black raven can actually *decrease* the probability that all ravens are black. Consider this example:

Suppose that there exists two worlds. The hypothesis we are considering is that all ravens in our world are black. In one of the existing worlds there are 100 black ravens, no non-black ravens, and 1,000,000 other birds of various types. In the other world there are 1,000 black ravens, 1 white raven, and 1,000,000 other birds of various types. A bird is now selected at random from one of the two worlds. It turns out to be a black raven. Just counting, we see that it is most likely that this black raven came from the second world—where not all ravens are black. Thus the selection of this black raven implies that there is a positive probability that there exists a white raven.

P. Maher [MAH] reasoned that, when it is highly probable that there are no ravens at all, then it is also highly probable that all ravens are black. He argues as follows:

Imagine an infinitely intelligent newborn baby having an innate ability to deal with formal logic, English syntax, and subjective probability. This baby proceeds to define a raven in detail. She then argues that it is extremely unlikely that there are any ravens. Therefore (vacuously) it is quite likely that all ravens are black. On the other hand, she reasons, if there do exist ravens, then there is a good chance that they come in a variety of colors. As a consequence of this reasoning, if she were to discover that a black raven existed, then she would consider assertion (9.5.1) to be less probable than was initially thought.

Maher has an accompanying argument that goes as follows. Consider a universe with exactly two objects in it, each of which is very unlikely to be a raven (a 1/1000 chance) and reasonably unlikely to be black (a 1/10 chance). It follows that the probability that all ravens are black is then decreased when it is discovered that one of the two objects is a black raven.

J. Hintikka changed the example a bit and came to some new conclusions. He considered the statement, "All men are tall." This is logically equivalent with "All short people are women." Thus, if a randomly selected person is a short woman, then this could be construed as evidence that all men are tall. Note that we lack background knowledge to indicate that there are dramatically fewer men than short people. Nonetheless, our intuition inclines us to reject the conclusion.

It has been suggested that "All ravens are black" and "All non-black things are non-ravens" can have different effects when they are *accepted*. The reasoning considers situations where the total numbers or prevalences of ravens and black objects are unknown, but are estimated. When the hypothesis "All ravens are black" is accepted, then it is argued that the estimated number black objects increases, while the estimated number of ravens does not change.

We can illustrate this last argument by considering two people who have identical information about ravens and black objects, and who have identical estimates of the numbers of ravens and of black objects that exist. Let us assume that there are 100 objects overall, and that each such object is just as likely to be a raven as a non-raven. Also each object is just as likely to be black as non-black. We further suppose that the statements "A is a raven" and "B is a raven" are independent for different objects A and B.

As a result, the estimated number of ravens is 50. And the estimated number of black things is 50. It follows that the estimated number of black ravens is 25. And the estimated number of non-black ravens is 25.

One of the two people involved performs a test of the hypothesis that "All ravens are black." The other person tests the hypothesis that "All non-black objects are non-ravens." To keep things simple, let us assume that the evidence used for these tests has nothing to do with the collection of 100 objects that we are dealing with here.

If the first person accepts the hypothesis that "All ravens are black," then about 50 objects whose colors were previously in doubt (i.e., the ravens) are now thought to be black; at the same time, nothing different is thought about the remaining objects (the non-ravens). As a consequence, the first person should estimate the number of black ravens as 50, the number of black non-ravens as 25, and the number of non-black non-ravens as 25. The argument we are presenting restricts the domain of "All ravens are black" to ravens.

On the other hand, if the second person accepts the hypothesis that "All non-black objects are non-ravens," then the approximately 50 non-black objects about which it was uncertain whether each was a raven will be thought to be non-ravens. Simultaneously, nothing different will be thought about the approximately 50 remaining objects (the black objects). As a result, this second person should estimate the number of black ravens at 25, the number of black non-ravens at 25, and the number of non-black non-ravens at 50. By this reasoning, since the two people disagree about their estimates after they have accepted the different hypotheses, we see that accepting "All ravens are black" is not equivalent to accepting "All non-black things are non-ravens." Accepting the first of these means estimating more things to be black; while accepting the second of these statements involves estimating more things to be non-raven. Thus we see that the former requires as evidence ravens that turn out to be black and the latter requires non-black things that turn out to be non-ravens.

Exercise: In more detail, Maher's argument considers a universe of exactly two objects, each of which is very unlikely to be a raven (a 1/1000 chance) and reasonably unlikely to be black (a 1/10 chance). He finds that the probability that all ravens are black decreases from 0.9985 to 0.8995 when it is discovered that one of the two objects is a black raven. Explain.

9.6   The Prisoner's Dilemma

This is a classic problem that was first formulated by Merrill Flood and Melvin Dresher at the RAND Corporation in 1950. It illustrates how and why two rational individuals might not cooperate, even if it is in their best interest to cooperate. It was Albert Tucker of Princeton University who formalized the game and gave it its name.

Here is a brief and succinct description of the Prisoner's Dilemma:

> Two members A and B of a criminal gang are arrested and imprisoned. Each prisoner is in solitary confinement with no means of communicating with other prisoners or anyone else. The prosecutor has insufficient evidence to convict either prisoner on the principal charges. The prosecutor hopes to get both men sentenced to a year in prison on a lesser charge. So the prosecutor offers each man a bargain. Each prisoner is given the option of **(a)** betraying the other by testifying that the other committed the crime or **(b)** cooperating with the other by remaining silent. The deal is this:

- If A and B each betray the other, then each of them will serve 2 years in prison.

- If A betrays B but B remains silent, then A will be set free and B will serve 3 years in prison. If B betrays A but A remains silent, then B will be set free and A will serve 3 years in prison.

- If both A and B remain silent, then both of them will serve just 1 year in prison.

It is understood that each prisoner will have no opportunity to reward or punish the other prisoner other than the prison sentence that he gets, and that their decisions will not affect their reputations in the future.

So what will the prisoners do? Clearly the best thing is for them both to remain silent. But each prisoner will not do that because he will be afraid that the other guy will rat him out. Therefore each prisoner will betray the other one. Thus each will serve 2 years (instead of the more desirable 1 year).

In reality, people tend to be more prone to be cooperative than the cold analysis in the last paragraph would suggest.

There are a number of variants of the Prisoner's Dilemma game. In one of these variants the process is iterated several times. This means that the prisoners each decide whether to betray or to remain silent. They are each told of the result. Then they are once again offered the deal (on top of the previous deal) and they must decide again how to proceed. This iteration process could involve two steps, or three steps, or any number of steps. Assuming that the players know in advance how many iterations there will be, what is the logical thing for the prisoners to do in this situation?

The answer is obtained by "backward induction." It is clear that, in the last iteration, each prisoner should logically betray the other prisoner. This is because each prisoner knows that his opponent will not have a chance to later retaliate. So both players will rat the other out in the last step. But then, in the preceding iteration, it is also clear that each prisoner's best strategy is to betray the other prisoner. Just because each player knows that his opponent will rat him out in the last step. And so on backward, step-by-step. Each prisoner betrays the other prisoner at each iteration.

The strategy just described is valid if the iteration has finitely many steps, whether the players know the number of steps or not. In case there are infinitely many iterations then the optimal strategy is not known.

The Prisoner's Dilemma can be used as a model for a number of real-world situations involving potentially cooperative behavior. The basic analysis may be applied to situations in which two entities could gain important benefits from cooperating or suffer from the failure to do so, but find it difficult or expensive to coordinate their activities to achieve cooperation.

Let us introduce some notation and make the entire game more analytical. If both prisoners remain silent, then they receive the reward R for cooperating. If each prisoner rats out the other one, then they both receive the punishment payoff P. If the first prisoner (call him A) betrays the second prisoner (call him B), while B remains silent, then A receives the "temptation" payoff T while B receives the "sucker's" payoff S. On the other hand, if A remains silent while B betrays A, then A receives the sucker's payoff S while B receives the temptation payoff T. We can exhibit this information in the following table. Note that, for this to be the Prisoner's Dilemma game in the strong sense, we must have the inequalities $T > R > P > S$ (compare with the verbal description of the Prisoner's Dilemma at the beginning of this section). We will use the words "cooperate" to mean "remains silent" and "defect" to mean "rats out the other prisoner."

Canonical PD Payoff Matrix

	Cooperate	Defect
Cooperate	R, R	S, T
Defect	T, S	P, P

The payoff relationship $R > P$ tells us that mutual cooperation is superior to mutual defection. But the payoff relationships $T > R$ and $P > S$ tell us that defection is the dominant strategy for both players. Game theorists would say that mutual defection is the only strong Nash equilibrium in the game (i.e., the only outcome from which each player could only do worse by unilaterally changing strategy).[1]

---

[1] Here John Nash was the mathematician who invented some of the modern ideas in game theory, including the famous Nash equilibrium, and who won the Nobel Prize for his work.

We may summarize the dilemma by saying that mutual cooperation yields a better outcome than mutual defection, but it is not the *rational outcome* because, from a self-interested perspective, the choice to cooperate (at the individual level) is irrational.

We may change the game a bit by considering a "donation game." In this setup, cooperation corresponds to offering the other player a benefit b at a personal cost of c. Here we assume that $b > c$. Defection in this context means offering nothing. The playoff matrix now becomes

New PD Payoff Matrix

	Cooperate	Defect
Cooperate	$b-c, b-c$	$-c, b$
Defect	$b, -c$	$0, 0$

Observe that $2(b-c) > b-c$, hence the donation game is qualified to be an iterated game.

The donation game may be applied to markets. Let us assume that A grows oranges and B grows apples. The marginal utility of an apple for the orange grower A is b. This is higher than the marginal utility c of an orange, since A has a surplus of oranges and no apples. Likewise, for apple-grower B, the marginal utility of an orange is b while the marginal utility of an apple is c. If A and B agree to exchange an apple and an orange, and each person is true to the deal, then each receives a payoff of $b-c$. If one of the players "defects" and does not deliver on the promise, then the defector will receive a payoff of b, while the cooperator will lose c. If both defect, then neither player gains or loses anything at all.

Exercise: Imagine a situation just like the original prisoner's dilemma described here, but with three prisoners instead of two. How should they behave?

9.7 The Eyes Have It

A certain island has people with two brown eyes and people with two blue eyes. There are no others. Nobody knows their own eye color, and there are no mirrors. The islanders *never* speak to each other about eye color.

According to the religion practiced on the island, if an islander learns of his/her own eye color, then that person must commit suicide the next day.

The apple cart is upset one day when a visiting politician from the mainland stands up to give a speech. Not familiar with the local customs, he comments that he himself has blue eyes, and he is charmed and surprised to see a resident in this part of the world who also has blue eyes. What are the consequences of this action?

For the sake of argument, let us suppose that just three of the islanders have blue eyes, and their names are A, B, and C. The islanders are all very smart and very logical. Islander A reasons that B can see at least one blue-eyed person (namely C). And B knows that C can see at least one blue-eyed person (namely A). But A does not know that B knows that (because A does not know his/her own eye color). So A reasons as follows: If I have brown eyes, then B can only see one blue-eyed person (namely C). And if B can only see one blue-eyed person, he/she will think, "If I have brown eyes, then C cannot see anybody with blue eyes." "Thus," thinks A, "if I have brown eyes, then B does not know whether C can see anyone with blue eyes. By saying, "I see a blue-eyed person," the foreign politician has removed this uncertainty. Thus A, B, and C will each conclude that they have blue eyes. Too bad for them.

This same reasoning can be applied to any number of people with blue eyes.

Here is an alternative, and slightly more sophisticated, approach to the problem. Suppose that there are n blue-eyed people on the island. We will perform mathematical induction on the index n. The statement that we shall establish is

> If there are n blue-eyed people on the island, then n days after the politician makes his speech, all n blue-eyed people must commit suicide. (9.7.1)

In case $n = 1$, the single, blue-eyed person must realize that the politician is referring to him or her when he says that he sees a blue-eyed islander (remember that this single, blue-eyed person can see that all the other people on the island have brown eyes). Thus he/she must die the next day.

Now suppose that the assertion has been established for $n = j$. Let us consider the situation where there are $j + 1$ people with blue eyes. Each blue-eyed person will reason as follows. "If I am not blue-eyed, then there are only j blue-eyed people on the island. So, by the inductive hypothesis (in the first sentence), they all must die after j days. But, when j days pass and none of them die, then the thinker realizes that there are not j blue-eyed people on the island but rather $j + 1$ blue-eyed people. And that $(j + 1)$st blue-eyed person is him/herself!

We have demonstrated that statement (9.7.1) is true for $n = 1$. And, if it is true for $n = j$, then it must be true for $n = j + 1$. It follows then that (9.7.1) is true for all values of n.[2]

Now here is a game involving hats that uses logic similar to that which we just used to analyze the eyes of the islanders. Suppose that there are three players, and a hat is placed on each of their heads. Each hat is either red or blue. Each player can see the other two player's hats, and determine their color. But no player can see his/her own hat. The problem is for each player to determine the color of the hat that he/she is wearing.

The players are told to raise their hands if they see a red hat. All three players raise their hands. After a moment's thought, one player announces "red" and wins. How did he/she do it?

If player 1 sees a blue hat on the head of player 2, then player 1 knows that he/she himself/herself is *not* wearing a blue hat; for, if he/she were, then player 3 would see only blue hats and would not have his/her hand raised. Thus any player who sees a blue hat immediately can infer that his/her own hat is red. We conclude therefore that none of the players is seeing a blue hat. So all the hats must be red.

Exercise: What would happen in the last example if only two players raised their hands?

---

[2]This reasoning is due to Terence Tao.

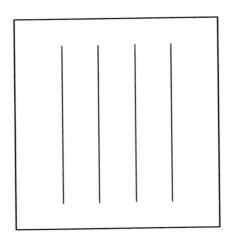

Figure 9.20: A sheet with four parallel lines.

9.8   A Sure Bet

You draw four parallel lines on a piece of paper (Figure 9.20). You fold the paper along the dotted line indicated in Figure 9.21.

You explain to a friend that you will, on one half of the paper, connect the lines in two pairs. [There are three different ways to do this; Figure 9.22 shows the three different things that you might do.] *You do not show the friend what you have done.* With your work face down on the table, you then invite the friend to do as you did: connect the four remaining loose ends in two pairs.

Then you place a bet with the friend: if the resulting figure, when the paper is unfolded, is a continuous loop then you win the bet; if the resulting figure is instead two disjoint loops, then the friend wins the bet.

The bet is even money. Is your friend wise to take this bet? This is another one of those situations that, on the surface, sounds eminently fair. It is not.

Examine Figure 9.23. With the paper unfolded, it shows all the different things you might have done (your work is always on the top) plotted against all the things that your friend might have tried (her work is always on the bottom).

Notice that in six of the nine possible configurations the result is a single continuous loop whereas in only three of the nine configurations does the

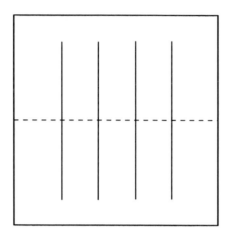

Figure 9.21: The sheet to be folded.

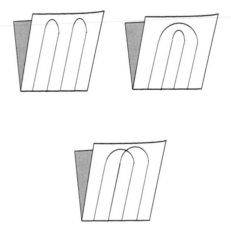

Figure 9.22: The folded sheet.

Figure 9.23: All possible configurations.

result consist of two disjoint loops. You have a 2/3 chance of winning the bet.

Your friend would be foolish to take this bet.

The fellow who contributed this problem says that he put his son through college by going to bars and betting people on the outcome of this game.

It is curious to note that you, as the perpetrator, do not even have to *know* anything to be the odds-on favorite to win this bet. *No matter what configuration you choose for your end of the paper*, you have a two out of three chance of ending up with a single closed loop and thus winning the bet. Figure 9.23 illustrates this last point.

Exercise: Imagine playing a game as in this section, but beginning with four lines as in Figure 9.24. Analyze this game and determine who would be the winner.

9.9   Hilbert's Hotel Infinity

In 1934 the eminent mathematician David Hilbert, in a public lecture, proposed the idea of "Hotel Infinity." It is a fascinating notion, and we describe it here. The charming and entertaining book [GAM] explores the idea further.

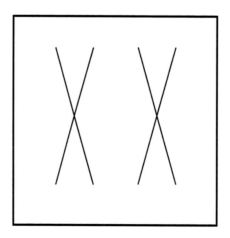

Figure 9.24: A new game.

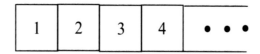

Figure 9.25: The hotel infinity.

Imagine a hotel with infinitely many rooms. For convenience, these rooms are numbered $1, 2, 3, \ldots$. See Figure 9.25.

Now suppose that, on a given evening, all the rooms are occupied. But a new guest, who does *not* have a reservation, shows up. What is the hotel manager to do?

If this were an ordinary hotel with finitely many rooms, then the manager would be stuck. She would have no room for the new guest, and would have to send her away. But instead she has infinitely many rooms, so she can do something creative:

> She moves the guest in Room 1 to Room 2, and the guest in Room 2 to Room 3, and the guest in Room 3 to Room 4, and so on. In this way every one of the pre-existing guests will still have a room. But Room 1 will be freed up and she can put the new guest in Room 1.

Miraculous, no??

The same thing would work if 5 new guests were to show up. Here is how she would proceed:

> She moves the guest in Room 1 to Room 6, and the guest in
> Room 2 to Room 7, and the guest in Room 3 to Room 8, and so
> on. In this way every one of the pre-existing guests will still have
> a room. But Rooms 1 through 5 will be freed up and she can put
> the new guests in Rooms 1 through 5.

Now what if infinitely many new guests show up? That is more of a
challenge. But the hotel manager is a talented mathematical thinker, and
she comes up with this solution:

> She moves the guest in Room 1 to Room 2. Then she moves the
> guest in Room 2 to Room 4. Next she moves the guest in Room 3
> to Room 6. In general, she moves the guest in Room k to Room
> $2k$. This frees up Rooms $1, 3, 5, 7, \ldots$. So infinitely many rooms
> have been freed up, and she can accommodate infinitely many
> new guests.

The devil is working to try to trip up our intrepid hotel manager. So he
comes up with the following situation. He brings in infinitely many buses,
and each bus has infinitely many passengers. He wants her to provide ac-
commodations for *all* of these new guests. How does she do it?

> First, she moves the guest in Room 1 to Room 2. Then she
> moves the guest in Room 2 to Room 4. Next she moves the guest
> in Room 3 to Room 6. And so on.
> This frees up all the odd-numbered rooms. But she needs to
> accommodate infinitely many collections of infinitely many peo-
> ple. How can she do it?
> She takes the people from the first bus and puts them in the
> rooms numbered
> $$3^1, 3^2, 3^3, \ldots$$

So the first bus passengers all go into rooms whose numbers are
powers of 3. These are of course odd-numbered rooms, so they
are vacant.

Next, she takes the people from the second bus and puts them in the rooms numbered

$$5^1 , 5^2 , 5^3 , \ldots$$

So the second bus passengers all go into rooms whose numbers are powers of 5. These are of course odd-numbered rooms, and they are different from the rooms with numbers that are powers of 3. So these rooms are vacant, and can accommodate the passengers from bus number 2.

For the next step, she takes the people from the third bus and puts them in the rooms numbered

$$7^1 , 7^2 , 7^3 , \ldots$$

So the third bus passengers all go into rooms whose numbers are powers of 7. These are of course odd-numbered rooms, and they are different from the rooms with numbers that are powers of 3 or powers of 5. So these rooms are vacant, and can accommodate the passengers from bus number 3.

The hotel manager continues in this fashion, placing passengers from bus k into rooms with numbers that are powers of the kth odd prime number.

Take particular note that, for the fourth bus, we do not put the guests into the rooms numbered $9^1 , 9^2 , 9^3 , \ldots$ Because these are actually powers of 3, so these rooms would have been filled when we accommodate the guests from the first bus. It is important that we only use rooms with numbers that are *powers of the prime numbers*. So, for the fourth bus, we use the rooms with numbers $11^1 , 11^2 , 11^3 , \ldots$.

The method that we have described for infinitely many buses each with infinitely many passengers certainly works, but it also leaves infinitely many rooms empty. For example, the room with number $15 = 5 \cdot 3$ will be empty. But this may be convenient for the hotel.

In fact Georg Cantor (1845–1918) was the man who developed techniques for handling infinite sets. He showed that infinite sets come in different sizes, and he created a calculus for manipulating infinite sets. He is remembered today as one of the great mathematicians of the last hundred years. Hilbert's

Hotel Infinity was in fact a commentary on and an appreciation of Cantor's work.

Exercise: Imagine infinitely many buses showing up, each with infinitely many passengers, but these people are only willing to stay in rooms with even numbers. How can the hotel manager accommodate these new guests?

Chapter 10

Breaking the Code

10.1 Alan Turing and Cryptography

10.1.1 Background on Alan Turing

Alan Mathison Turing was born in 1912 in London, England. He died tragically in 1954 in Wilmslow, Cheshire, England. Today, Turing is considered
to have been one of the great mathematical minds of the twentieth century.
He did not invent cryptography (as we shall see, even Julius Caesar engaged
in cryptography). But he ushered cryptography into the modern age. The
current vigorous interaction of cryptography with computer science owes its
genesis in significant part to the work of Turing. Turing also played a decisive
role in many of the key ideas of modern logic. It is arguable that Turing had
the key ideas for inventing the stored program computer (although it was
John von Neumann (1903–1957), another twentieth-century mathematical
genius, who together with Herman Goldstine (1913–2004), actually carried
out the ideas).

Turing had a tragically short life, marked by disappointments and frustra-
tions. Today he is considered to be one of the great geniuses of mathematical
thought. His story is recounted in [TUR] and [HOD].

Turing had difficulty fitting in at the British "public schools" that he
attended. [Note that a "public school" in Britain is what we in America
would call a private school.] Young Turing was more interested in pursuing

his own thoughts than in applying himself to the dreary school tasks that were designed for average students. At the Sherborne School, Turing had little patience for the tedious math techniques that the teachers taught. Yet he won almost every mathematics prize at the school. He was given poor marks in penmanship, and he struggled with English.

Turing had a passion for science beginning at a very young age. He later said that the book *Natural Wonders Every Child Should Know* by Edwin Tenney Brewster had a seminal influence on him. When he was still quite young, he read Einstein's papers on relativity and he read Arthur Eddington's account of quantum mechanics in the book *The Nature of the Physical World.*

In 1928, at the Sherborne School, Alan Turing became friends with Christopher Morcom. Now he had someone in whom he could confide, and with whom he could share scientific ideas and inquiries. Turing had never derived such intellectual companionship from either his classmates or his rather diffident schoolteachers. Sadly, Morcom died suddenly in 1930. This event had a shattering effect on the young Alan Turing. The loss of his companion led Turing to consider spiritual matters, and over time this led him to an interest in physics.

It may be mentioned that Turing early on developed an interest in sports. He was a very talented athlete—almost at the Olympic level—and he particularly excelled at running. He maintained an interest in sports throughout his life.

In 1931 Alan Turing entered King's College at Cambridge University. Turing earned a distinguished degree at King's in 1934, followed by a fellowship at King's. In 1936 he won the Smith Prize for his work in probability theory. In particular, Turing was one of the independent discoverers of the Central Limit Theorem.

In 1935 Turing took a course from Max Newman on the foundations of mathematics. Thus his scientific interests took an abrupt shift. The hot ideas of the time were Gödel's incompleteness theorem—which says that virtually any mathematical theory will have true statements in it that cannot be proved—and (what is closely related) David Hilbert's questions about decidability.

In 1936 Alan Turing published his seminal paper "On Computable Numbers, with an Application to the Entscheidungsproblem." Here the *Entscheidungsproblem* is the fundamental question of how to decide—in a manner

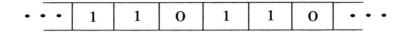

Figure 10.1: A deterministic Turing machine.

that can be executed by a machine—when a given mathematical question is provable. In this paper Turing first described his idea for what has now become known as the *Turing machine*. We now take a mathematical detour to talk about Turing machines.

10.1.2   The Turing Machine

A *Turing machine* is a device for performing effectively computable operations. It consists of a machine through which a bi-infinite paper tape is fed. The tape is divided into an infinite sequence of congruent boxes (Figure 10.1). Each box has either a numeral 0 or a numeral 1 in it. The Turing machine has finitely many "states" $S_1, S_2, \ldots, S_n$. In any given state of the Turing machine, one of the boxes is being scanned.

After scanning the designated box, the Turing machine does one of three things:

(1) It erases the numeral 1 that appears in the scanned box and replaces it with a 0, or it erases the numeral 0 that appears in the scanned box and replaces it with a 1, or it leaves the box unchanged.

(2) It moves the tape one box (or one unit) to the left or to the right.

(3) It goes from its current state S_j into a new state S_k.

It turns out that every logical procedure, every algorithm, every mathematical proof, every computer program can be realized as a Turing machine. The Turing machine is a "universal logical device." The next section contains a simple instance of a Turing machine. In effect, Turing had designed a computer before technology had made it possible to actually build one.

An Example of a Turing Machine

Here is an example of a Turing machine for calculating $x + y$:

State	Old Value	New Value	Move (l. or r.)	New State	Explanation
0	1	1	R	0	pass over x
0	0	1	R	1	fill gap
1	1	1	R	1	pass over y
1	0	0	L	2	end of y
2	1	0	L	3	erase a 1
3	1	0	L	4	erase another 1
4	1	1	L	4	back up
4	0	0	R	5	halt

If you look hard at the logic of this Turing machine, you will see that it thinks of x as a certain number of 1s, and it thinks of y as a certain number of 1s. It scans the x units, and writes a 1 to the right of these; then it scans y units, and writes a 1 to the right of these. The two blocks of 1s are joined into a single block (by erasing the space in between) and then the two extra 1s are erased. The result is $x + y$. As an exercise, you should provide the details of this argument in simple English sentences.

10.1.3   What Is Cryptography?

We use Alan Turing's contributions as a touchstone for our study of cryptography. Cryptography is currently a very hot field, due in part to the availability of high-speed digital computers to carry out decryption algorithms, in part to new and exciting connections between cryptography and number theory and logic, and in part to the need for practical coding methods both in industry and in government.

The discussion of cryptography that appears below is inspired by the book [KOB]. We refer the reader to that source for additional ideas and further reading.

As we always do in mathematics, let us begin by introducing some terminology. Cryptography is the study of methods for sending text messages in disguised form in such a manner that only the intended recipient can remove the disguise and read the message. The original message that we wish to send is called the *plaintext* and the disguised message is called the *ciphertext*. We shall always assume that both our plaintext and our ciphertext are written in the standard roman alphabet (i.e., the letters A through Z) together perhaps with some additional symbols like "blank space (⊔)", "question mark (?)", and so forth. The process of translating a plaintext message into a ciphertext message is called *encoding* or *enciphering* or *encrypting*. The process of translating an encoded message back to a plaintext message is called *deciphering* or sometimes *de-encrypting*.

For convenience, we usually break up both the plaintext message and the ciphertext message into blocks or units of characters. We call these pieces the *message units*, but we may think of them as "words" (however they are not necessarily English words). Sometimes we will declare in advance that all units are just single letters, or perhaps pairs of letters (these are called *digraphs*) or sometimes triples of letters (called *trigraphs*). Other times we will let the units be of varying sizes—just as the words in any body of text have varying sizes. An *enciphering transformation* is a function that assigns to each plaintext unit a ciphertext unit. The *deciphering transformation* is the inverse mapping that recovers the plaintext unit from the ciphertext unit. Any setup as we have just described is called a *cryptosystem*.

In general it is awkward to mathematically manipulate the letters of the alphabet. We have no notions of addition or multiplication on these letters. So it is convenient to associate to each letter a number. Then we can manipulate the numbers. For instance, it will be convenient to make the assignment

$$
\begin{aligned}
A &\leftrightarrow 0 \\
B &\leftrightarrow 1 \\
C &\leftrightarrow 2 \\
&\cdots \\
X &\leftrightarrow 23 \\
Y &\leftrightarrow 24 \\
Z &\leftrightarrow 25 .
\end{aligned}
$$

Thus if we see the message

$$22\ 7\ 0\ 19\ 12\ 4\ 22\ 14\ 17\ 17\ 24$$

then we can immediately translate this to

WHATMEWORRY

or

WHAT ME WORRY?

Notice that, in cryptography, we generally do not worry about capital and lowercase letters. Everything is uppercase. Quite often we will not have a symbol for "blank space," so a little extra thought must be given at the end (as in this last example) to extract the message from the decrypted sequence of letters.

One device of which we will make frequent and consistent use is *modular arithmetic*. Recall that if n and k are integers, then $n \bmod k$ is that unique integer n' between 0 and $k - 1$ inclusive such that $n - n'$ is divisible by k. For example,

$$13 \bmod 5 = 3\,,$$

$$-23 \bmod 7 = 5\,,$$

$$82 \bmod 14 = 12\,,$$

$$10 \bmod 3 = 1\,.$$

How do we calculate these values? Look at the first of these. To determine $13 \bmod 5$, we *divide* 5 into 13: Of course 5 goes into 13 with quotient 2 and remainder 3. It is the *remainder* that we seek. Thus

$$13 \bmod 5 = 3\,.$$

It is similar with the other examples. To determine $82 \bmod 14$, divide 14 into 82. It goes 5 times with remainder 12. Hence

$$82 \bmod 14 = 12\,.$$

It is convenient that modular arithmetic respects the arithmetic operations. For example,

$$8 \times 7 = 56 \qquad \text{and} \qquad 56 \bmod 6 = 2\,.$$

But

$$8 \bmod 6 = 2 \quad \text{and} \quad 7 \bmod 6 = 1 \quad \text{and} \quad 2 \times 1 = 2\,.$$

So it does not matter whether we pass to mod 6 *before* multiplying or *after* *multiplying*. Either way we obtain the same result 2. Similar properties hold for addition and subtraction. One must be a bit more cautious with division, as we shall see below.

We supply some further examples:

$$[3 \bmod 5] \times [8 \bmod 5] = 24 \bmod 5 = 4 \,;$$

$$[7 \bmod 9] + [5 \bmod 9] = 12 \bmod 9 = 3 \,;$$

$$[4 \bmod 11] - [9 \bmod 11] = -5 \bmod 11 = 6 \,.$$

Now we begin to learn some cryptography by way of examples.

EXAMPLE 10.1 We use the ordinary 26-letter Roman alphabet A–Z, with the numbers 0–25 assigned to the letters as indicated above. Let $S = \{0, 1, 2, \ldots, 25\}$. We will consider units consisting of single letters. Thus our cryptosystem will consist of a function $f : S \to S$ which assigns to each unit of plaintext a new unit of ciphertext. In particular, let us consider the specific example

$$f(P) = \begin{cases} P + 5 & \text{if } P < 21 \,, \\ P - 21 & \text{if } P \geq 21 \,. \end{cases}$$

Put in other words,

$$f(P) = (P + 5) \bmod 26 \,. \tag{10.1.1}$$

Next let us use this cryptosystem to encode the message

GO AWAY

or

GOAWAY.

The first step is that we transliterate the letters into numbers (because, as noted earlier, numbers are easier to manipulate). Thus GOAWAY becomes 6 14 0 22 0 24.

Now we apply the "shift encryption" (10.1.1) to this sequence of numbers. Notice that

$$f(6) = (6 + 5) \bmod 26 = 11 \bmod 26 = 11 \,,$$

$$f(14) = (14 + 5) \bmod 26 = 19 \,,$$

$$f(0) = (0 + 5) \bmod 26 = 5 \,,$$

$$f(22) = (22 + 5) \bmod 26 = 1 \,,$$

$$f(0) = (0 + 5) \bmod 26 = 5 \,,$$

$$f(24) = (24 + 5) \bmod 26 = 3 \,.$$

Thus our ciphertext is 11 19 5 1 5 3. In practice, we may convert this ciphertext back to roman letters using our standard correspondence ($A \leftrightarrow 0$, $B \leftrightarrow 1$, etc.). The result is LTFBFD. Thus the encryption of "GO AWAY" is "LTFBFD." Notice that we have no coding for a blank space, so we ignore it.

This is a very simple example of a cryptosystem. It is said that Julius Caesar used this system with 26 letters and a shift of 3. We call this encryption system a "shift transformation."

Now let us use this same cryptosystem to encode the word "BRAVO." First, we translate our plaintext word to numbers:

$$1 \ 17 \ 0 \ 21 \ 14 \,.$$

Now we add 5 mod 26 to each numerical entry. The result is

$$6 \ 22 \ 5 \ 0 \ 19 \,.$$

Notice that the fourth entry is 0 because

$$21 + 5 \bmod 26 = 26 \bmod 26 = 0 \bmod 26 \,.$$

Thus if we wanted to send the message "BRAVO" in encrypted form, we would send 6 22 5 0 19. We can translate the encrypted message to roman letters as "GWFAT."

Conversely, we *decrypt* a message by subtracting 5 mod 26. Suppose, for instance, that you receive the encrypted message

$$\text{YMFSPDTZ}$$

Of course we first transliterate the message (using our usual correspondence) to a sequence of numbers:

$$24 \ 12 \ 5 \ 18 \ 15 \ 3 \ 19 \ 25 \, .$$

We decrypt by applying the function $f^{-1}(Q) = (Q - 5) \bmod 26$. The result is

$$19 \ 7 \ 0 \ 13 \ 10 \ 24 \ 14 \ 20 \, .$$

This easily translates to

THANKYOU

or

THANK YOU.

In a typical, real-life circumstance, you receive an encrypted message and *you do not know the method of encryption*. It is your job to figure out how to decode the message. We call this process *breaking the code*, and the science of codebreaking is called *cryptoanalysis*.

If the codebreaker happens to know that the message he/she has received is encrypted using a shift transformation, then there is a reasonable method to proceed. Imagine that you receive the message

CQNKNJCUNBOXANENA

Looks like nonsense. But the cryptographer has reason to believe that this message has been encoded using a shift transformation on single letters of the 26-letter alphabet. It remains to find the numerical value of the shift.

We use a method called *frequency analysis*. The idea of this technique is that it is known that "E" is the most frequently occurring letter in the English language. Thus we may suppose that the most frequently occurring character in the ciphertext is the encryption of "E" (*not* "E" itself). In fact we see that the character "N" occurs five times in the ciphertext, and that is certainly the most frequently occurring letter. If we hypothesize that "N" is the encryption of "E", then we see that "4" has been translated to "13" in the encryption. Thus the encryption key is $P \mapsto (P+9) \bmod 26$. And therefore the decryption scheme is $P \mapsto (P-9) \bmod 26$. If this putative decryption scheme gives a sensible message, then it is likely the correct choice (as any other decryption scheme will likely give nonsense). Let us try this scheme and see what result it gives. We have

$$\text{CQNKNJCUNBOXANENA}$$

has numerical realization

$$2 \ \ 16 \ \ 13 \ \ 10 \ \ 13 \ \ 9 \ \ 2 \ \ 20 \ \ 13 \ \ 1 \ \ 14 \ \ 23 \ \ 0 \ \ 13 \ \ 4 \ \ 13 \ \ 0 \ \ .$$

Under our decryption scheme, this translates to

$$19 \ \ 7 \ \ 4 \ \ 1 \ \ 4 \ \ 0 \ \ 19 \ \ 11 \ \ 4 \ \ 18 \ \ 5 \ \ 14 \ \ 17 \ \ 4 \ \ 21 \ \ 4 \ \ 17$$

which has textual realization

$$\text{THEBEATLESFOREVER}.$$

In other words, the secret message is

$$\text{THE BEATLES FOREVER}.$$

The trouble with the shift transformation is that it is just too simple-minded. It is too easy to break. There are variants that make it slightly more sophisticated. For example, suppose that the East Coast and the West Coast branches of National Widget Corporation cook up a system for sending secret messages back and forth. They will use a shift transformation, but *in each week of the year they will use a different shift*. This adds a level of complexity to the process. But the fact remains that, using a frequency analysis, the code can likely be broken in any given week.

10.1.4 Encryption by Way of Affine Transformations

We can add a genuine level of sophistication to the encryption process by adding some new mathematics. Instead of considering a simple shift of the form $P \mapsto P + b$ for some fixed integer b, we instead consider an *affine transformation* of the form $P \mapsto aP + b$. Now we are both multiplying (or dilating) the element P by an integer a and then translating it by b.

Division in Modular Arithmetic

There is a subtlety in the application of the affine transformation method that we must consider before we can look at an example. If the encryption scheme is $P \mapsto Q \equiv aP + b$, then the decryption scheme must be the inverse function. In other words, we solve for P in terms of Q. This just involves elementary algebra, and we find that

$$P = [1/a](Q - b) \bmod 26 .$$

We see that decryption, in the context of an affine transformation, involves division in arithmetic modulo 26. This is a new idea, and we should look at a couple of simple examples before we proceed with our cryptographic considerations.

We want to consider division modulo 26. Thus if a and b are whole numbers, then we want to calculate b/a and we want the answer to be another whole number modulo 26. This is possible only because we are cancelling multiples of 26, and it will only work when a has no common prime factors with 26. Let us consider some examples.

First let us calculate $4/7 \bmod 26$. What does this mean? We are dividing the whole number 4 by the whole number 7, and this looks like a fraction. But things are a bit different in modular arithmetic. We seek a number k such that

$$\frac{4}{7} \bmod 26 = k$$

or

$$4 = 7 \cdot k \bmod 26$$

or

$$4 - 7 \cdot k \quad \text{is divisible by } 26 .$$

We simply try different values for k, and we find with $k = 8$ that

$$4 - 7 \cdot 8 = 4 - 56 = -52 \quad \text{is indeed divisible by } 26.$$

In conclusion,

$$\frac{4}{7} \bmod 26 = 8.$$

We see the somewhat surprising conclusion that the fraction $4/7$ can be realized as a whole number in arithmetic modulo 26.

Next let us try to calculate $1/4 \bmod 26$. This is doomed to fail, because 4 and 26 have the prime factor 2 in common. We seek an integer k such that

$$1 = 4 \cdot k \bmod 26,$$

or in other words

$$1 - 4k \quad \text{is a multiple of } 26.$$

But of course $4k$ will always be even, so $1 - 4k$ will always be odd—*it cannot be a multiple of the even number 26.* This division problem cannot be solved.

We conclude this brief discussion with the example $2/9 \bmod 26$. We invite the reader to discover that the answer is $6 \bmod 26$.

There is in fact a mathematical device for performing division in modular arithmetic. It is the classical Euclidean algorithm. This simple idea is one of the most powerful in all of number theory. It says this: If n and d are integers, then d divides into n some whole number q times with some remainder r, and $0 \leq r < d$. In other words,

$$n = d \cdot q + r.$$

You have been using this idea all your life when you calculate a long division problem (not using a calculator, of course). We shall see in the next example that the Euclidean algorithm is a device for organizing information so that we can directly perform long division in modular arithmetic.

EXAMPLE 10.2 Let us calculate $1/20$ in arithmetic mod 57. We apply the Euclidean algorithm to 57 and 20. Thus we begin with

$$57 = 2 \cdot 20 + 17 \,.$$

We continue by repeatedly applying the Euclidean algorithm to divide the divisor by the remainder:

$$
\begin{aligned}
20 &= 1 \cdot 17 + 3 \\
17 &= 5 \cdot 3 + 2 \\
3 &= 1 \cdot 2 + 1
\end{aligned}
$$

Now, as previously indicated, we utilize this Euclidean algorithm information to organize our calculations. Begin with the last line to write

$$
\begin{aligned}
1 &= 3 - 1 \cdot 2 \\
&= 3 - 1 \cdot (17 - 5 \cdot 3) \\
&= [20 - 17] - 1 \cdot ([57 - 2 \cdot 20] + 5 \cdot [20 - 17]) \\
&= 20 \cdot 8 + 17 \cdot (-6) - 57 \\
&= 20 \cdot 8 + (57 - 2 \cdot 20) \cdot (-6) - 57 \\
&= 20 \cdot 20 - 7 \cdot 57 \,.
\end{aligned}
$$

This calculation tells us that $1 = 20 \cdot 20 \bmod 57$. In other words, $1/20 = 20 \bmod 57$.

For You to Try: Calculate $1/25 \bmod 64$ using the Euclidean algorithm.

Instances of the Affine Transformation Encryption

EXAMPLE 10.3 Let us encrypt the message "GO AWAY" using the affine transformation $P \mapsto 5P + 6 \bmod 26$. As usual,

GO AWAY   has numerical realization 6 14 0 22 0 24.

Under the affine transformation, we obtain the new numerical realization

10 24 6 12 6 22.

In roman letters, the message becomes the ciphertext

$$\text{KYGMGW.}$$

In order to decrypt the message, we must use the *inverse* affine transformation. If $R = 5P + 6 \bmod 26$, then $P = [1/5](R - 6) \bmod 26$. Using modular arithmetic, we see that 10 corresponds to

$$[1/5](10 - 6) = [1/5] \cdot 4 = 6 \bmod 26$$

(because $5 \cdot 6 \bmod 26 = 30 \bmod 26 = 4 \bmod 26$). Likewise 24 corresponds to

$$[1/5](24 - 6) = [1/5] \cdot 18 = 14 \bmod 26$$

(because $5 \cdot 14 \bmod 26 = 70 \bmod 26 = 18 \bmod 26$). We calculate the rest of the correspondences:

$$[1/5](6 - 6) = [1/5] \cdot 0 = 0 \bmod 26$$

(because $5 \cdot 0 \bmod 26 = 0 \bmod 26$). Next,

$$[1/5](12 - 6) = [1/5] \cdot 6 = 22 \bmod 26$$

(because $5 \cdot 22 \bmod 26 = 110 \bmod 26 = 6 \bmod 26$). Again,

$$[1/5](6 - 6) = [1/5] \cdot 0 = 0 \bmod 26 \, .$$

And, finally,

$$[1/5](22 - 6) = [1/5] \cdot 16 = 24 \bmod 26$$

(because $5 \cdot 24 \bmod 26 = 16 \bmod 26$).

In sum, we have applied our decryption algorithm to recover the message

$$6 \ \ 14 \ \ 0 \ \ 22 \ \ 0 \ \ 24.$$

This transliterates to

$$\text{GOAWAY}$$

or

GO AWAY.

In a real-life situation—if we were attempting to decrypt a message—we would not know in advance which affine transformation was used for the encoding. We now give an example to illustrate how to deal with such an eventuality.

EXAMPLE 10.4 We continue to work with the 26-letter Roman alphabet. We receive a block of ciphertext and wish to decode it. We notice that the most frequently occurring character in the ciphertext is "M" and the second most frequently occurring character in the ciphertext is "R." It is well known that, in ordinary English, the most commonly occurring letter is "E" and the second most commonly occurring letter is "T." So it is natural to hypothesize that we are dealing with an affine transformation that assigns "E" to "M" and "T" to "R."

This means that we seek an affine transformation $f(P) = aP + b$ such that $f(4) = 12 \bmod 26$ and $f(19) = 17 \bmod 26$. All arithmetic is, as usual, modulo 26. We are led then to the equations

$$12 \; = \; a \cdot 4 + b \bmod 26 \,,$$
$$17 \; = \; a \cdot 19 + b \bmod 26 \,.$$

We subtract these two equations to eliminate b and obtain

$$-5 = a \cdot (-15) \bmod 26$$

or

$$a = [-5/(-15)] \bmod 26 \,.$$

The solution is $a = 9$. Substituting this value into the first equation gives $b = -24 = 2 \bmod 26$.

Thus our affine encoding transformation is (we hope)

$$f(P) = 9P + 2 \,.$$

It is also easy to determine that the inverse (or decoding) transformation is $f^{-1}(Q) = [Q - 2]/9$.

For You to Try: Use the affine decryption scheme in the last example to
decode the message "ZMDEMRILMRRMZ."

Next we present an example in which an expanded alphabet is used.

Consider the standard Roman alphabet of 26 characters along with the
additional characters "blank space" (denoted ␣), "question mark" (?), "period (.)", and "exclamation point (!)". So now we have 30 characters, and
arithmetic will be module 30. As usual, we assign a positive integer to each
of our characters. Thus we have

$$
\begin{aligned}
A &\leftrightarrow 0 \\
B &\leftrightarrow 1 \\
C &\leftrightarrow 2 \\
&\ \cdots \\
X &\leftrightarrow 23 \\
Y &\leftrightarrow 24 \\
Z &\leftrightarrow 25 \\
␣ &\leftrightarrow 26 \\
? &\leftrightarrow 27 \\
. &\leftrightarrow 28 \\
! &\leftrightarrow 29
\end{aligned}
$$

Because there are now 30 different characters, we also use 30 different numerical codes—the numbers from 0 to 29.

Imagine that we receive a block of ciphertext, and that we wish to decode
it. We notice that the most commonly used characters in the ciphertext are
"D" and "!". It is known that the most commonly used characters in ordinary
English are "␣" and "E."[1] If we assume that the ciphertext was encrypted
with an affine transformation, then we seek an affine mapping $f(P) = aP+B$
such that $f(␣) = D$ and $f(E) =!$. Thus we are led to $f(26) = 3$ and $f(4) = 29$

---

[1] We formerly said that "E" was the most commonly used letter. But that was before
we added the blank space ␣ to our alphabet.

and then to the system of equations

$$3 = a \cdot 26 + b \bmod 30,$$
$$29 = a \cdot 4 + b \bmod 30.$$

As before, we subtract the equations to eliminate b. The result is

$$-26 = 22a \bmod 30.$$

This equation is equivalent (dividing by 2) to

$$-13 = 11a \bmod 30.$$

Since 11 and 30 have no factors in common, we may easily find the unique solution $a = 7$. Substituting this value in the second equation gives $b = 1$. We conclude that our affine transformation is $f(P) = 7P + 1$.

If the ciphertext we have received is

$$21 \ \ 7 \ \ 29 \ \ 3 \ \ 14 \ \ 29 \ \ 12 \ \ 14 \ \ 7 \ \ 14 \ \ 19 \ \ 18 \ \ 29 \ \ 24,$$

then we can apply $f^{-1}(Q) = [Q - 1]/7$ to obtain the plaintext message

$$20 \ \ 18 \ \ 4 \ \ 26 \ \ 19 \ \ 4 \ \ 23 \ \ 19 \ \ 26 \ \ 18 \ \ 19 \ \ 24 \ \ 11 \ \ 4 \ \ 29.$$

This transliterates to

USE TEXT STYLE!

A nice feature of this example is that the spaces and the punctuation are built into our system of characters. Hence the translated message is quite clear, and requires no further massaging.

10.1.5   Digraph Transformations

Just to give an indication of how cryptographers think, we shall now consider digraphs. Instead of thinking of our message units as single characters, we will now have units that are *pairs* of characters. Put in other words, the plaintext message is broken up into two-character segments or words. [It should be stressed that these will not, in general, correspond to English words. Certainly words from the English language are generally longer than two letters. Here, when we say "word," we simply mean a unit of information.]

In case the plaintext message has an odd number of characters, then of course we cannot break it up evenly into units of two characters. In this instance we add a "dummy" character like "X" to the end of the message so that an even number of characters will result. Any English message will still be readable if an "X" is tacked on the end.

Let K be the number of elements in our alphabet (in earlier examples, we have seen alphabets with 26 characters and also alphabets with 30 characters). Suppose now that MN is a digraph (i.e., an ordered pair of characters from our alphabet). Let x be the numerical equivalent of M and let y be the numerical equivalent of N. Then we assign to the digraph MN the number $x \cdot K + y$. Roughly speaking, we are now working in base-K arithmetic.

EXAMPLE 10.5 Let us work in the familiar Roman alphabet of 26 characters. A common digraph in English is "TH." Notice that the numerical equivalent of "T" is 19 and the numerical equivalent of "H" is 7. According to our scheme, we assign to this digraph the single number $19 \cdot 26 + 7 = 501$.

It is not difficult to see that each positive integer corresponds to a unique digraph. Consider the number 358. Then 26 divides into 358 a total of 13 times with a remainder of 20. We conclude that 358 corresponds to the digraph with numerical equivalents 13 20. This is the digraph "NU."

It is straightforward to see that the greatest integer that can arise in this labeling scheme for digraphs is for the digraph $\Omega\Omega$, where Ω is the last character in our alphabet. If the first character is assigned to 0 (as we have done in the past) then the last character is assigned to $K - 1$ (where K is the number of characters in the alphabet). The numerical label is then $(K - 1) \cdot K + (K - 1) = K \cdot K - 1$. So it is safe to say that $K^2 - 1$ is an upper bound for numerical labels in our digraph system.

We conclude, then, that an enciphering transformation is a function that consists of a rearrangement of the integers $\{0, 1, 2, \ldots, K^2 - 1\}$. One of the simplest such transformations is an *affine transformation* on $\{0, 1, 2, \ldots, K^2 - 1\}$. We think of this set of integers as $\mathbb{Z}$ modulo K^2. So the encryption has the form $f(P) = aP + b \bmod K^2$. As usual, the integer a must have no prime factors in common with K^2 (and hence no prime factors in common with K).

EXAMPLE 10.6 We work as usual with the 26-letter Roman alphabet. There are then 26×26 digraphs, and these are enumerated by means of the integers $0, 1, 2, \ldots, 26^2 - 1$. In other words, we work in arithmetic modulo 676, where of course $676 = 26^2$. The digraph "ME" has letters "M" corresponding to "12" and "E" corresponding to "4." Thus we assign the digraph number $12 \cdot 26 + 4 = 316 \bmod 676$.

If our affine enciphering transformation is $f(P) = 97 \cdot P + 230$, then the digraph "ME" is encrypted as $97 \cdot 316 + 230 = 462 \bmod 676$.

If instead we consider the digraph "EM" then we assign the integer $4 \cdot 26 + 12 = 116$. And now the encryption is $97 \cdot 116 + 230 = 666 \bmod 676$.

EXAMPLE 10.7 Suppose that we want to break a digraphic encryption system that uses an affine transformation. So we need to determine a and b. This will require two pieces of information.

Let us attempt a frequency analysis. From statistical studies, it is known that some of the most common digraphs are "TH," "HE," and "EA." The most common ones that include the "blank space" character are "E␣," "S␣," and "␣T." If we examine a good-sized block of ciphertext and notice the most commonly occurring digraphs, then we might suppose that those are the encryptions of "TH" or "HE" or "EA." Consider for example the ciphertext (based on the 27-character alphabet consisting of the usual 26 letters of the Roman alphabet plus the blank space, and numbered 0 through 26)

$$\text{XIHZYIQHRCZJSDXIDCYIQHPS} \quad .$$

We notice that the digraphs "XI," "YI," and "QH" each occur twice in the message. We might suppose that one of these is the encryption of "TH," one is the encryption of "HE," and one is the encryption of "EA" (although, as indicated above, there are other possibilities). Let us attempt to directly solve for the affine transformation that will decrypt our ciphertext. The affine transformation will have the form $f^{-1}(Q) = a'Q + b'$ and our job is to find a' and b'.

To be specific, let us guess that

$$\text{TH} \quad \text{encrypts as} \quad \text{YI}$$

$$\text{HE} \quad \text{encrypts as} \quad \text{XI} \, .$$

This means that we have the numerical correspondences

$$520 \leftrightarrow 656$$

and

$$193 \leftrightarrow 629 \, .$$

So we have the algebraic equations

$$520 = a' \cdot 656 + b' \bmod 729 \, ,$$
$$193 = a' \cdot 629 + b' \bmod 729 \, .$$

Subtracting the equations as usual (to eliminate b'), we see that

$$327 = a' \cdot 27 \bmod 729 \, .$$

Unfortunately this equation does not have a unique solution, because 27 and 729 have prime factors in common (such as 3).

We make another guess. Let us suppose that

$$\text{TH encrypts as QH}$$

$$\text{HE encrypts as YI} \, .$$

This means that we have the numerical correspondences

$$520 \leftrightarrow 439$$

and

$$193 \leftrightarrow 656 \, .$$

So we have the algebraic equations

$$520 = a' \cdot 439 + b' \bmod 729$$
$$193 = a' \cdot 656 + b' \bmod 729 \, .$$

Subtracting the equations as usual (to eliminate b'), we see that

$$327 = a' \cdot 217 \bmod 729 \, .$$

Now 217 and 729 have no prime factors in common, so we may solve for a' uniquely. The answer is $a' = 408$. Substituting into our first equation gives $b' = 13$. So our decryption algorithm is

$$f^{-1}(Q) = 408Q + 13 \, . \tag{10.1.2}$$

We apply this rule to the ciphertext

XIHZYIQHRCZJSDXIDCYIQHPS.

For example, the digraph "XI" has numerical equivalent 629. It translates, with decryption rule (10.1.2), to 37. This in turn corresponds to the plaintext digraph "BK." We can already tell we are in trouble, because there is no word in the English language that contains the two letters "BK" in sequence.

It is our job then to try all the other possible correspondences of encrypted digraphs "XI," "YI," and "QH" to the plaintext digraphs. We shall not work them all out here. It turns out that the one that does the trick is

$$XI \text{ is the encryption of } TH$$

and

$$QH \text{ is the encryption of } EA.$$

Let us try it and see that it successfully decrypts our secret message.

The proposed correspondences have numerical interpretation

$$629 \leftrightarrow 520$$

and

$$439 \leftrightarrow 108.$$

This leads to the equations

$$520 = a' \cdot 629 + b' \bmod 729,$$
$$108 = a' \cdot 439 + b' \bmod 729.$$

Subtracting as usual, we obtain

$$412 = a' \cdot 190 \bmod 729.$$

Since 190 and 729 have no prime factors in common, we can certainly divide by 190 and solve for a'. We find that $a' = 547$. Substituting into the second equation gives $b' = 545$. In conclusion, the decrypting transformation is $f^{-1}(Q) = 547Q + 545 \bmod 729$.

Now we can systematically apply this affine transformation to the digraphs in the ciphertext and recover the original message. Let us begin:

$$XI \to 629 \xrightarrow{f^{-1}} 520 \to TH,$$

$$HZ \to 214 \xrightarrow{f^{-1}} 234 \to IS.$$

The calculations continue, and the end result is the original plaintext message

THIS HEART OR THAT HEADX

As you can see, an "X" is affixed to the end to force the message to have an even number of characters (counting blank spaces) so that the digraph method will work.

One important point that the last example illustrates is that cryptography will always entail a certain amount of (organized) guesswork.

Exercise: Encrypt the message

Trump is President.

using an affine encryption scheme. Now figure out how to decrypt it.

10.2   RSA Encryption

10.2.1   Basics and Background

Modern security considerations make it desirable for us to have new types of encryption schemes. In the old days (beginning even with Julius Caesar), it was enough to have a method for disguising the message that we were sending. For example, imagine that the alphabet is turned into numeric symbols by way of the scheme

$$A \longmapsto 0$$
$$B \longmapsto 1$$
$$C \longmapsto 2$$

and so forth.

Then use the encryption

$$n \longmapsto n + 3 \bmod 26. \tag{10.2.1}$$

And now convert these numbers back to roman letters. As a simple example, the phrase

WHAT ME WORRY

translates to the string of integers

$$22 \quad 7 \quad 0 \quad 19 \quad 12 \quad 4 \quad 22 \quad 14 \quad 17 \quad 17 \quad 24.$$

The encryption (10.2.1) turns this string of integers into

$$25 \quad 10 \quad 3 \quad 22 \quad 15 \quad 7 \quad 25 \quad 17 \quad 20 \quad 20 \quad 2$$

and this, in turn, transliterates to

ZKDWPHZRUUC.

It is easy to imagine that there would be scenarios in which the following could occur:

- You wish to have a means that a minimum-wage security guard (whom you don't necessarily trust) can check that people entering a facility know a password—but you don't want him to know the password.

- You wish to have a technology that allows anyone to encrypt a message—using a standard, published methodology—but only someone with special additional information can decrypt it.

- You wish to have a method to be able to convince someone else that you can perform a procedure, or solve a problem, or prove a theorem, without actually revealing the details of the process.

This may all sound rather dreamy, but in fact—thanks to the efforts and ideas of R. Rivest, A. Shamir, and L. Adleman—it is now possible. The so-called RSA encryption scheme is now widely used. For example, the e-mail messages that I receive on my cell phone are encrypted using RSA.

In this discussion, we shall describe how RSA encryption works, and we shall encrypt a message using the methodology.

10.2.2   Preparation for RSA

We now sketch the background ideas for RSA. These are all elementary ideas from basic mathematics. It is remarkable that these are all that are needed to make this profound new idea work.

Computational Complexity

• Suppose that you have a deck of N playing cards and you toss them in the air. Now you want to put them back into their standard order. How many "steps" will this take? [We want to answer this question in such a manner that a machine could follow the instructions.] First we look through all N cards and find the first card in the ordering.

Then we look through the remaining $N - 1$ cards and find the second card in the ordering.

And so forth.

So the re-ordering of the cards takes

$$N + (N - 1) + (N - 2) + \cdots 3 + 2 + 1 = \frac{N(N + 1)}{2}$$

steps. [Refer to Section 2.7 for more details of this example.] Notice that the answer here is a quadratic polynomial in N. Thus we say that the problem can be solved in polynomial time.

• We have heard a rumor that the four-color theorem is true. So we have a graph with N vertices and we wish to color each vertex, using either red, yellow, blue, or green. See Figure 10.2. The only rule is that two adjacent vertices (i.e., vertices that are connected by a segment or edge) cannot be the same color.

Of course the number of possible colorings is the number of functions from the set with N objects to the set with 4 objects. That is 4^N. [Again refer to Section 2.7.] The machine, being as dumb as it is, will simply try

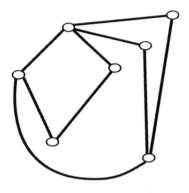
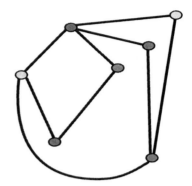

A typical planar graph.   An admissible coloring for
                               the planar graph.

Figure 10.2: A graph that we wish to color.

all the possible colorings until it finds one that works. Thus we see that
the number of steps is now an exponential function of N. We call this an
exponential time problem.

10.2.3   Modular Arithmetic

This is a familiar idea, and we have already alluded to it earlier in this
chapter. The "right" way to define the idea is with cosets, but we shall
content ourselves here with a more informal definition.

When we write $n \mod k$, we mean simply the remainder when n is divided by k. Thus

$$25 = 1 \quad \mod 3\,,$$

$$15 = 3 \quad \mod 4\,,$$

$$-13 = -3 \quad \mod 5 = 2 \quad \mod 5\,.$$

It is an important fact—which again is most clearly seen using the theory
of cosets—that modular arithmetic respects sums and products. We shall use
these facts in a decisive manner below.

Fermat's Theorem

Let a and b be two (positive) integers. We say that a and b are relatively prime if they have no common prime factors. For example,

$$72 = 2^3 \cdot 3^2$$

$$175 = 5^2 \cdot 7$$

hence 72 and 175 are relatively prime.

If n is an integer, let $\mathcal{P}(n)$ be the set of integers less than n that are relatively prime to it. Let $\varphi(n)$ be the number of elements in $\mathcal{P}(n)$.

> **Theorem:** If n is a positive integer and k is relatively prime to n then
> $$k^{\varphi(n)} = 1 \quad \mod n.$$

The proof of this result is easy if you have some modern algebra under your belt. The idea is this. The collection $\mathcal{P}(n)$ of numbers relatively prime to n forms a group under multiplication. This means that we may multiply any two members of $\mathcal{P}(n)$ and obtain another element of $\mathcal{P}(n)$. Also each element of $\mathcal{P}(n)$ has a multiplicative inverse in $\mathcal{P}(n)$. Thus any element of the group, raised to the power $\varphi(n)$ (the number of elements in the group) will equal 1 modulo n.

For later use, it is worth noting that if p, q are prime numbers and $n = p \cdot q$, then $\varphi(n) = (p-1) \cdot (q-1)$.

The reason is that the only numbers less than or equal to n that are not relatively prime to n are $p, 2p, 3p, \ldots q \cdot p$ and $q, 2q, 3q, \cdots (p-1)q$.

There are q numbers in the first list and $p-1$ numbers in the second list. The set $\mathcal{P}(n)$ of numbers relatively prime to n is the complement of these two lists, and it therefore has

$$pq - q - (p-1) = pq - q - p + 1 = (p-1) \cdot (q-1) \equiv \varphi(n)$$

elements.

10.2.4   Relatively Prime Integers

Two integers a and b are relatively prime if they have no prime factors in common. As noted above, for example, 72 and 175 are relatively prime.

It is a fundamental fact of elementary number theory that if a, b are relatively prime, then we can find other integers x and y such that

$$xa + yb = 1. \qquad (10.2.1)$$

For example, we have noted that $a = 72$ and $b = 175$ are relatively prime. The corresponding integers x, y are $x = -17$ and $y = 7$. Thus

$$(-17) \cdot 72 + 7 \cdot 175 = 1.$$

One can prove this result using Fermat's theorem above. For, since b is relatively prime to a, thus

$$b^{\varphi(a)} = 1 \quad \bmod a.$$

But this just says that

$$b^{\varphi(a)} - 1 = k \cdot a$$

for some integer k. Unraveling this equation gives (10.2.1).

In practice, one finds x and y using the Euclidean algorithm (otherwise known as long division).

In the example of 72, 175, one calculates:

$$
\begin{aligned}
175 &= 2 \cdot 72 + 31 \\
72 &= 2 \cdot 31 + 10 \\
31 &= 3 \cdot 10 + 1.
\end{aligned}
$$

You know you are finished when the remainder is 1. For now we have

$$
\begin{aligned}
1 &= 31 - 3 \cdot 10 \\
&= 31 - 3 \cdot (72 - 2 \cdot 31) \\
&= 7 \cdot 31 - 3 \cdot 72 \\
&= 7 \cdot (175 - 2 \cdot 72) - 3 \cdot 72 \\
&= 7 \cdot 175 - 17 \cdot 72.
\end{aligned}
$$

That is the decomposition we seek.

10.2.5   The RSA System Enunciated

Now we can quickly and efficiently describe how to implement the RSA encryption system, and we can explain how it works.

Imagine that, when he was president, George W. Bush had an important
message that he wished to send to Donald Rumsfeld. Of course Rumsfeld
was a highly placed man of many responsibilities, and you can imagine that
Bush's message was quite secret. So he wanted to encode the message:

Your time is up. Hasta la vista, baby.

So Bush goes to the library and finds the RSA encryption book. This is
a readily available book that anyone can access. It is not secret. A typical
page in the book reads like this:

NAME	VALUE OF n	VALUE OF e
Puck, Wolfgang	4431 ... 7765	8893 ... 4453
Rehnquist, William	6667 ... 2345	1235 ... 9877
Riddle, Nelson	7583 ... 2391	4637 ... 4389
Rin Tin-Tin	5355 ... 5353	5465 ... 7647
Rogers, Roy	7859 ... 4359	3053 ... 2935
Roosevelt, Theodore	7835 ... 2523	7893 ... 4231
Rotten, Johnny	3955 ... 4343	4485 ... 9921
Roy, Rob	3795 ... 5441	2219 ... 3319
Rumsfeld, Donald	1117 ... 8853	9265 ... 2383
Russert, Tim	6463 ... 4647	3223 ... 3231
Simpson, Orenthal James	6677 ... 2235	4857 ... 2223

What does this information mean? Of course we know, thanks to Euclid,
that there are infinitely many primes. So we can find prime numbers with
as many digits as we wish. Each number n in the RSA encryption book is

the product of two 75-digit primes p and q: Thus $n = p \cdot q$. Each number e is chosen to be a number with at least 100 digits that is relatively prime to $\varphi(n) = (p-1) \cdot (q-1)$. Of course we do not publish the prime factorization of the number n; we also do not publish $\varphi(n)$. All that we publish is n and e for each individual. Now an important point to understand is that Bush does not need to understand any mathematics or any of the theory of RSA encryption in order to encode his message. [Well, it would be nice if he understood modular arithmetic. But he is, after all, the President of the United States.] All he does is this:

- First he breaks the message into units of five letters. We call these "words," even though they may not be English language words.

For the message from Bush to Rumsfeld, the "words" would be

$$\text{YOURT} \quad \text{IMEIS} \quad \text{UPHAS} \quad \text{TALAV} \quad \text{ISTAB}$$
$$\text{ABY}$$

- He transliterates each "word" into a sequence of numerical digits, using our usual scheme of translation.

- Then he encodes each transliterated word w with the rule

$$w \longmapsto w^e \mod n .$$

Bush will send to Rumsfeld this sequence of encrypted words. That is all there is to it.

The real question now is:

What does it take to decrypt the encoded message? How can Rumsfeld read the message?

This is where some mathematics comes into the picture. We must use Fermat's theorem, and we must use our ideas about relatively prime integers. But the short answer to the question is this. If $\widetilde{w}$ is a word encrypted according to the simple scheme described above, then we decrypt it with this algorithm:

> We find integers x and y so that $xe + y\varphi(n) = 1$ and then we calculate
> $$\widetilde{w}^x \quad \text{mod } n.$$

That will give the decrypted word w with which we began. Since w has only five characters, and n has 150 characters, we know that $w \mod n = w$—so there is no ambiguity arising from modular arithmetic. We can translate w back into roman characters, and we recover our message.

Now here is the most important point in our development thus far:

> In order to *encrypt* a message, we need only look up n and e in the public record RSA encryption book. But, in order to decrypt the message, we must know x. Calculating x necessitates knowing $\varphi(n)$, and that necessitates knowing the prime factorization of n.

It is a theorem that calculating the prime factorization of an integer with k digits is a problem of exponential complexity in k. For an integer with 150 digits, using a reasonably fast computer, it would take quite a long time to find the prime factorization.

10.2.6   The RSA Encryption System Explicated

In fact, with all the preliminary setup that we have in place, it is a simple matter to explain the RSA encryption system.

For suppose that we selected an $n = p \cdot q$ and an e relatively prime to $\varphi(n) = (p - 1) \cdot (q - 1)$ corresponding to a particular person listed in the RSA encryption book. If we are the certified decrypter, then we know the prime factorization of n—that is, we know that $n = p \cdot q$ for specific prime numbers p and q.

We therefore know $\varphi(n) = p \cdot q - p - q + 1 = (p-1) \cdot (q-1)$ and so we can calculate the x and the y in the identity $xe + y\varphi(n) = 1$. Once we know x, then we know everything. For

$$
\begin{aligned}
\widetilde{w}^x \mod n &= \left[w^e\right]^x \mod n \\
&= w^{ex} \mod n \\
&= w^{1-y\varphi(n)} \mod n \\
&= w \cdot \left[w^{\varphi(n)}\right]^{-y} \mod n \\
&= w \cdot 1^{-y} \mod n \\
&= w \mod n \\
&= w
\end{aligned}
$$

since w is certainly relatively prime to n.

This shows how we recover the original word w from the encrypted word $\widetilde{w} = w^e \mod n$.

Exercise: Explain how RSA encryption can be used so that a minimally paid security guard can check that each person entering a facility knows a certain password but *he* does not know the password.

Chapter 11

Discrete Problems

11.1 Far-Reaching Dominoes

You have an unlimited supply of dominoes that measure $1'' \times 2''$. You are working in a room that is 10 feet long and has no ceiling. You begin at one wall, place a domino on the floor, place another atop it (not necessarily squarely), and proceed to pile dominoes (Figure 11.1). Will your pile be able to reach the other wall, 10 feet away, without toppling?

The key physical observation (which we take for granted) is that, if the jth domino protrudes a distance of λ_j inches beyond the end of the $(j-1)^{\text{th}}$ domino, then the moment of inertia[1] of the jth domino is

$$\rho \cdot (\lambda_j)^2/2.$$

Here ρ is the linear density of the domino. For simplicity, we assume that ρ is 1. Then the moment of inertia of the jth domino is $(\lambda_j)^2/2$.

If we stack N dominoes in the manner indicated, then the total moment

---

[1]The moment of inertia, otherwise known as the angular mass or rotational inertia, of a rigid body is a tensor that determines the torque needed for a desired angular acceleration about a rotational axis. It depends on the body's mass distribution and the axis chosen, with larger moments requiring more torque to change the body's rotation. It is an extensive (additive) property: the moment of inertia of a composite system is the sum of the moments of inertia of its component subsystems (all taken about the same axis).

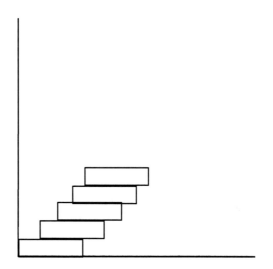

Figure 11.1: Stacking of dominoes.

of inertia of the system is

$$M = \rho \cdot \sum_{j=2}^{\infty} \frac{\lambda_j^2}{2}.$$

Here the summation notation on the right means

$$\frac{\lambda_2^2}{2} + \frac{\lambda_3^2}{2} + \frac{\lambda_4^2}{2} + \cdots.$$

[Notice that we start the sum with $j = 2$ since the first domino lies flat on the ground and has no moment that is relevant to this problem.]

Assume that c is a positive constant. Suppose that we take $\lambda_2 = c/2$, $\lambda_3 = c/3$, and, in general, $\lambda_j = c/j$. Then

$$M = \rho \cdot \sum_{j=2}^{N} \frac{(c/j)^2}{2} = \frac{c^2}{2} \sum_{j=2}^{N} \frac{1}{j^2}.$$

It is known that this sum converges to a finite number that depends on the parameter c but is independent of N.[2] However the sum of the lengths of

---

[2]If you are not familiar with series, then you should get out your calculator or computer and add up the first 50, then the first 100, then the first 1000 terms of this series so that you can learn its behavior.

the protrusions is

$$L = \sum_{j=2}^{N} \lambda_j = \sum_{j=2}^{N} \frac{c}{j} = c \cdot \sum_{j=2}^{N} \frac{1}{j}.$$

This sum becomes large without bound as N becomes larger and larger.[3]

If c is a very small positive number, then we see that the moment of inertia is as small as we please; we can certainly choose c small enough that the stack will not tumble. But, since the sum defining L becomes large without bound, the stack of dominoes will reach arbitrarily far to the right.

The answer to the problem is "yes," the stack of dominoes *can* reach 10 feet to the far end of the room.

Challenge: [This is Tricky] Find an estimate for how many dominoes will be needed to reach across the room, as indicated in the problem. [*Hint:* The number required is enormous. You may wish to use your computer as an experimental tool.]

Exercise: Suppose now that the dominoes are shaped like equilateral triangles. How does that change the answer to the question?

11.2   Surreal Life

The first number system that any child learns about is the *natural numbers*. These are the numbers

$$\mathbb{N} = \{1, 2, 3, \ldots\}.$$

They are the counting numbers, and a child begins his/her study of arithmetic by counting things.

Next in school the child is taught about fractions. These are numbers of the form

$$\frac{p}{q}, \quad p, q \in \mathbb{N}, \ q \neq 0.$$

---

[3]Again use your calculator or computer to sum up terms of this series. You will need a *lot* of terms to see that this sum becomes arbitrarily large. Try 1 million terms, then 10 million terms. Then 100 million terms.

With the fractional number system the child comes to grip with fractions, with dividing a whole into parts, with the concept of sharing.

Later on, the child learns about negative numbers. This is a bit more abstract, and requires some effort to master. In particular, the young student learns about the *integers*

$$\mathbb{Z} = \{\pm n : n \in \mathbb{N}\} \cup \{0\}$$

and the *rational number system*

$$\mathbb{Q} = \left\{ \pm \frac{p}{q} : p, q \in \mathbb{N}, q \neq 0 \right\}.$$

When the student gets to the stage of studying serious mathematics, he/she learns about trigonometry and the theory of functions, then he/she is finally introduced to the *real numbers*. For our purposes it is most convenient to think of the real numbers as the set of all possible numbers that are represented by decimal expansions. The real numbers includes the rational numbers (numbers with terminating or nonterminating and repeating decimal expansions) and the irrational numbers (numbers with nonterminating, nonrepeating decimal expansions).

Most people in everyday life get by with the rational numbers only. When you go to a hardware store and order lumber, you specify the dimensions using rational numbers. When you go to the grocery store and order flour, you specify the weight using rational numbers. But mathematicians really need the real number system. Just as an instance, the Pythagorean theorem tells us that the diagonal of a square of side 1 has length $\sqrt{2}$, and $\sqrt{2}$ is irrational. As a second example, the circumference of a circle of radius 1 is 2π, and 2π is irrational. Advanced mathematics entails many different limiting processes, and we need to have a number system in which all our limits exist. The real number system fills that bill. The rational number system does not.

Engineers and physicists are also content to work with the real number system. They occasionally refer to the complex number system, but the real numbers are their bread and butter. All of the mathematical operations that any mathematical scientist might care to perform are safe to carry out in the real number system. The answer will always be some new real number.

But as mathematicians we are naturally curious, and we may wonder whether there was a "larger" number system out there. The mathematicians who originally formulated calculus (Newton and Leibniz) spoke of "infinitesimal numbers." These are positive numbers that are (i) not equal to zero but (ii) smaller than any ordinary positive number. For more than 200 years scientists wrestled with the idea of infinitesimal. Conceptually and intuitively speaking, infinitesimals are quite appealing. But nobody knew whether they really existed. Then, in 1966, Abraham Robinson [ROB] constructed a new number system, called the *nonstandard real numbers*, in which infinitesimals exist rigorously. Somewhat later, in 1976, John Horton Conway came up with another way to approach the matter. He constructed the *surreal numbers*.

What we would like to do here is to give you an idea of what the surreal numbers are. This will be your first experience of "constructing" a number system, and you should find it fascinating and appealing. For more information about the surreal numbers, consult [CON]. It may be noted that Conway himself rather humbly just called these objects *numbers*. It was the noted computer scientist Donald Knuth [KNU3] who named them surreal numbers. Knuth's book actually predates Conway's. The surreal numbers are perhaps the only significant mathematical discovery to be first publicized in a novel.

Here we shall follow Conway's exposition rather closely. Let L and R be two sets of numbers. Assume that no member of L is greater than or equal to any member of R. Then $\{L|R\}$ is a surreal number. All surreal numbers are constructed in this fashion.

If the surreal number $x = \{L|R\}$, then we write x^L as a typical element of L and x^R as a typical element of R. We will sometimes then write $x = \{x^L|x^R\}$.

If we write $x = \{a, b, c, \ldots | d, e, f, \ldots\}$, then we are saying that $x = \{L|R\}$ where $a, b, c, \ldots$ are typical elements of L and $d, e, f, \ldots$ are typical elements of R.

Now we need to define some binary operations and relations on the surreal numbers.

- Let us say what it means to write $x \geq y$ for x and y surreal numbers. Write $x = \{x^L|x^R\}$ and $y = \{y^L|y^R\}$. Then $x \geq y$ provided that every x^R is greater than y and x is greater than every y^L. Do not get confused here. Of course x and y are surreal numbers. But, in general, x^L, y^L, x^R, and y^R are also surreal numbers. So it makes sense to compare x^R with y and x with y^L. We shall be able to make the idea of $\geq$ clearer with the examples that we present below.

- We say that $x = y$ if $x \geq y$ and $y \geq x$.

- We say that $x > y$ if $x \geq y$ and $y \not\geq x$.

- We define
$$x + y = \{x^L + y, x + y^L | x^R + y, x + y^R\}\,.$$

- We define
$$-x = \{-x^R | -x^L\}\,.$$

- We define
$$
\begin{aligned}
x \cdot y \; = \; \{ & x^L y + xy^L - x^L y^L, x^R y + xy^R - x^R y^R | x^L y + xy^R \\
& -x^L y^R, x^R y + xy^L - x^r y^L\}\,.
\end{aligned}
$$

That is the end of our basic description of the surreal numbers. The remarkable fact is that this set, as we have defined it, forms a field. Basically this means that we have operations of addition and multiplication that are commutative and associative, that there is an additive identity and a multiplicative identity, and that each nonzero surreal number has a multiplicative inverse. This field contains the real numbers (as we shall see), and it contains many new numbers and objects as well.

Let us begin our discussion by considering some examples of surreal numbers. Recall that a surreal number has the form $\{L|R\}$, where L and R are two sets of numbers that we have constructed earlier. Now the most primitive set that we know is the empty set $\emptyset$—the set with no elements. So we may consider the surreal number $\{\emptyset|\emptyset\}$. Note that this *is* a legitimate surreal number because every element of L is less than every element of R (because both sets are empty, so the assertion is vacuous). We usually denote this surreal number a bit more elegantly by $\{\ |\ \}$, and we call it 0. And in fact this surreal number will turn out to be the additive identity. That is to say, if x is any surreal number, then $x + 0 = 0 + x = x$. We shall say more about this matter below.

It is routine (and tedious) to verify certain standard properties of 0. We shall do one of them now. Notice that it is not the case that $0^R \leq 0$ or $0 \leq 0^L$. Therefore $0 \leq 0$. We may also easily see that $-0 = 0$.

Next, we can use the empty set $\{ \ \}$ and the set $\{0\}$ to form our next surreal numbers. We may consider $\{ \ | \ \}$, $\{0|0\}$, $\{0| \ \}$, and $\{ \ |0\}$. The first of these we discussed in the preceding paragraph, and it is the surreal number 0. The second of these is *not* a legitimate surreal number since $0 \leq 0$. The third item *is* a legitimate surreal number, and we call it 1. This number 1 will turn out to be the multiplicative identity. In other words, for any surreal number x, $1 \cdot x = x \cdot 1 = x$. The last item in our list is also a legitimate surreal number, and we call it -1. It is worth noting that our new number -1 is indeed the negative of the new number 1—according to the fifth bulleted definition at the beginning of this discussion.

We may now ask some silly questions. For instance, is $1 \leq 0$? The answer is "no," because there is a 1^L with $0 \leq 1^L$. Similar reasoning shows that it *is* the case that $0 \leq 1$. Also $-1 < 0$.

We may go on, if we wish, and define

$$\{1| \ \} = 2 \ , \ \{0|1\} = 1/2 \ , \ \{ \ |-1\} = -2 \ , \ \{-1|0\} = -1/2 \, .$$

One may simply check the definitions to see that

$$-2 < -1 < -\frac{1}{2} < 0 < \frac{1}{2} < 1 < 2 \, .$$

One may also calculate that $1 + 1 = 2$.

In a similar vein, we may define $3 = 1 + 2 = \{2| \ \}$, $4 = 1 + 3 = \{3| \ \}$, and so forth. So we now have all the positive integers. We may further define $1/4 = \{0|1/2\}$ (which we may check by calculating that $1/4 + 1/4 = 1/2$), $1/8 = \{0|1/4\}$, and so on. So now we have arbitrarily small positive numbers.

So far, nothing very exciting has happened. We have not seen any new numbers. We have instead just seen a new and somewhat arcane method of looking at old numbers. But now let us now consider the surreal number $\epsilon = \{0|1, 1/2, 1/4/1/8, \ldots\}$. And also look at the surreal number $\omega = \{1, 2, 4, 8, 16, \ldots| \ \}$. It turns out that ϵ is nonzero (obviously), is positive, and is smaller than $1, 1/2, 1/4, 1/8, \ldots$. Likewise, ω is nonzero, positive, and is larger than $1, 2, 4, 8, \ldots$. It can further be calculated that $\epsilon \cdot \omega = 1$. The surreal number ϵ is an example of a number that we think of as an *infinitesimal*. The surreal number ω is an example of a number that we think of as an *infinitary number*.

So now we see that the surreal numbers contains some new animals. And there are many more, including the ordinal numbers of Georg Cantor (1845–1918). We cannot delve into all the details here, but instead refer you to [CON] for further reading.

Exercise: Explain how multiplication and division would work in the surreal numbers.

11.3   A Problem with Marriage

In order to describe and motivate this problem, let us be a bit old fashioned and simple minded. And not politically correct, perhaps. We describe a social situation, which coincides with the way that many people remember American life to have been sixty years ago.

A young man comes of age. His aim is to get married. In seeking a wife, he decides to date at most 100 women. After he dates a woman for a while, he must either marry her, or reject her and move on to another. Once a woman is rejected, he may not return to her later. Ultimately he must choose just one woman and marry her.

The interesting feature of the problem is that the young man may look back, but he may not look ahead. At any point he may say "my current lady friend is more appealing, and we are more compatible together, than any woman that I've dated previously." On that basis he may decide to marry her. Or he may think "this woman is terrific, and more appealing than any of my previous dates, but I am going to gamble that someone better will come along soon."

The "marriage problem" is to determine the best strategy for the young man just described. In order to remove the emotional claptrap from the problem, we reformulate it as follows:

A hat is filled with 100 slips of paper. On each slip of paper is written some positive integer (note that *any* positive integer may appear on the slips—not merely the integers from 1 to 100). The integers do not necessarily appear in any sequence or pattern. Each of the slips has a different integer on it, so there is just one slip with the *greatest* integer.

A person, who has no prior knowledge of which numbers appear on the slips—but who does know that there are 100 slips—is to blindly pull slips from the hat one by one. The person looks at each slip, then either agrees to accept that number of dollars and quit the game or decides to go on and choose another slip.

Note that the contestant looks at each slip as he/she proceeds, and then decides whether to quit or to go on. Once the contestant sets a slip aside, he/she cannot return to it. He/she can go forward, but *cannot* go back. If no choice is made by the time the 100th slip is reached, then the contestant must accept the number of dollars on the 100th slip.

What is the best strategy for the contestant? [Here "best strategy" means that the contestant will have the optimum likelihood of garnering the greatest number of dollars.]

This author is embarrassed to admit that, when he was first told this problem, he thought a minute and said "Well, there is no strategy. It's hopeless." In part, the difficulty was that he didn't understand what a strategy was. The main problem was that he wasn't thinking.

We suppose in advance that a "strategy" will take the following form: the contestant will draw out a certain number of slips—say k of them—and make careful note of the numbers recorded. After the k slips have been drawn, the contestant will determine to select the next slip that satisfies Property P, where Property P is to be determined. We shall discuss later why it is reasonable to concentrate on this type of strategy.

Solution: We call the first slip that is drawn "slip 1," the next one drawn "slip 2," and so forth.

Our goal is to optimize the number of dollars received. Any strategy that would result in choosing the $(k+1)^{\text{st}}$ slip can be improved by remembering the largest recorded number M on the slips $1, 2, \ldots, k$ and then selecting the next slip that comes up that has a number greater than M (if no such slip ever comes up then the contestant is stuck with the 100th slip). Applying this observation over and over again, we find that the best strategy, given the parameters set in the paragraph before this solution began, is to take note of the largest number M on any of the slips $1, 2, \ldots, k$ and then select the next slip that has a number exceeding M.

This scheme having been established, our job is then to choose the best possible k. Suppose that the largest overall number Q (of all 100 slips) appears on slip $r + 1$. The contestant will not be successful in selecting that slip unless two conditions are fulfilled:

1) $r \geq k$ (because we are going to reject the first k slips, so if $r < k$, then the $(r + 1)^{\text{st}}$ slip, bearing the highest number, gets rejected).

2) The highest number on slips number 1 through r is also the highest number on slips 1 through k (for if the highest number P on slips 1 through r is *greater* than the highest number M on slips 1 through k, then $M < P$ and P will get chosen before the $(r + 1)^{\text{st}}$ slip is ever reached).

The probability that the overall largest number Q is on slip $r + 1$ (or on *any particular slip*, for that matter) is $1/100$. The probability of finding the slip with number Q on it, assuming that is slip number $r + 1$, is k/r (think, for example, of what can go wrong if $r = k + 1$). In sum, the probability of winning the game with the slip having largest number Q, given that that slip is the $(r + 1)^{\text{st}}$ and that we will reject the first r and choose the $(r + 1)^{\text{st}}$, is

$$p_r = \frac{1}{100} \cdot \frac{k}{r}.$$

The allowable values for r are $r = k, k + 1, \ldots, 99$. Thus the probability of winning the game, using the designated strategy, is

$$P = \sum_{r=k}^{99} p_r = \frac{k}{100} \sum_{r=k}^{99} \frac{1}{r}. \tag{11.3.1}$$

But now here is an important lesson about the sum on the right-hand side of the last formula:

If x is small and positive, then we may write

$$\ln(1 + x) = x \cdot \left\{ \ln\left[(1 + x)^{1/x} \right] \right\}$$

(we discussed the logarithm function in Section 2.1) and the expression in brackets inside the logarithm on the right is the expression that we use to define Euler's number $e \approx 2.718\ldots$ as $x \to 0$. Thus we have that

$$\ln(1 + x) \approx x \cdot \ln e = x.$$

We apply this observation to our sum as follows:

$$
\begin{aligned}
\ln N &= \ln\left[\frac{N}{N-1} \cdot \frac{N-1}{N-2} \cdots \frac{3}{2} \cdot \frac{2}{1}\right] \\
&= \ln\left(\frac{N}{N-1}\right) + \ln\left(\frac{N-1}{N-2}\right) + \cdots + \ln\left(\frac{3}{2}\right) + \ln\left(\frac{2}{1}\right) \\
&= \ln\left(1 + \frac{1}{N-1}\right) + \ln\left(1 + \frac{1}{N-2}\right) \\
&\qquad + \cdots + \ln\left(1 + \frac{1}{2}\right) + \ln\left(1 + \frac{1}{1}\right).
\end{aligned}
$$

Now we may apply our observation that $\ln(1 + x) \approx x$ to each of these summands, with the role of $x > 0$ being played by $1/(N-1)$, $1/(N-2)$, ... The result is that

$$\ln N \approx \frac{1}{N-1} + \frac{1}{N-2} + \cdots + \frac{1}{2} + \frac{1}{1}.$$

[Check this approximation formula on your calculator or computer to see how accurate it is!] Thus we see that

$$\sum_{r=k}^{99} \frac{1}{r} = \sum_{r=1}^{99} \frac{1}{r} - \sum_{r=1}^{k-1} \frac{1}{r} \approx \ln 99 - \ln(k-1) = \ln\left(\frac{99}{k-1}\right).$$

But, with this approximation, the formula (11.3.1) for the probability of getting the slip with the biggest number by passing up k slips and then picking the succeeding slip with the next largest number that we see is

$$P \approx \frac{k}{100} \cdot \ln\left(\frac{99}{k-1}\right).$$

We wish to select k so that this probability is as large as possible.

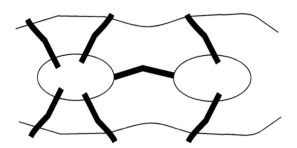

Figure 11.2: The seven bridges of Königsberg.

Of course calculus is custom made for maximizing a function of this sort. But this book does not presuppose that the reader knows any calculus. Instead, take out your graphing calculator or computer algebra software and, by inspecting the graph of the function

$$P(x) = \frac{x}{100} \cdot \ln\left(\frac{99}{x-1}\right),$$

determine where P takes its largest value. You will find that the answer is approximately at $x = 100/e$, where e (as previously noted) is Euler's number $e \approx 2.718\ldots$.

We conclude from this analysis that the player should examine the first $100/e$ slips (rounded off to the nearest whole number, which is 37) that he/she draws, remembering the largest number observed on these slips. The next slip that comes up that has a number larger than that observed maximum is the one to choose. This is the optimal strategy.

Exercise: Suppose that the young man had to choose two wives. How would that alter his strategy?

11.4   Euler's Bridges

The village of Königsberg is currently part of Russia, but in Euler's day (Leonhard Euler, 1707–1783) it was part of Prussia. The town is notable because it has two large islands, and these are connected to the mainland and to each other by seven bridges. See Figure 11.2.

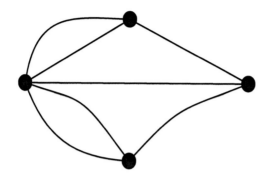

Figure 11.3: The graph representing the seven bridges of Königsberg.

The figure shows the land mass and the two islands. The seven bridges are also clearly depicted by heavy segments. It had been a problem of some standing to determine whether it was possible to walk a path that crossed each bridge exactly once. The only rule here is that, once a bridge is entered, one must walk all the way across it. And one cannot walk back. You can only cross each bridge once.

Euler's profound insight was that one should construct from this problem a graph. Here we are not talking about the graph of a function but rather about a combinatorial graph. One accomplishes this task by placing a vertex or node in each land mass and then connecting two vertices with an edge if the two corresponding land masses are connected by a bridge. See Figure 11.3. Now the problem becomes

Can one traverse the graph shown in Figure 11.3 along a path in such a fashion that each edge is traveled once and only once?

What Euler noticed was that, except for the node where the path begins and the node where the path ends, each of the nodes is entered and exited the same number of times. One does not stop at a node. One passes through it. Therefore, of the four nodes, three of them must be the terminus of an even number of edges. But such is not the case. In fact all four of the nodes has odd valence—meaning that an odd number of edges terminates there. Thus the desired path is impossible.

The problem changes if we demand that the path begin and end at the same place. For then that beginning/ending node will also have even valence. In fact all the nodes will have to have even valence. But such is not the case.

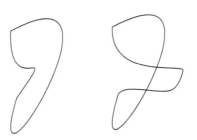

A simple, closed curve.   Not a simple closed curve.

Figure 11.4: A simple, closed curve.

The Königsberg bridge circuit is impossible.

Now let us look at a different type of geometry problem. We write γ : $[0, 1] \to \mathbb{R}^2$ for a closed curve that does not cross itself. This notation means that γ is a function, its domain is the interval $[0, 1] \equiv \{x \in \mathbb{R} : 0 \leq x \leq 1\}$, and its range is the set of ordered pairs of real numbers. We are assuming that $\gamma(0) = \gamma(1)$ but $\gamma(s) \neq \gamma(t)$ otherwise. This is commonly called a *simple, closed curve*. See Figure 11.4.

Prove that there are four points A, B, C, D lying on the curve such that these four points are the vertices of a rectangle.

Look at Figure 11.5. At the left is a quadrilateral with vertices on the given curve. At the right there is also a quadrilateral with vertices on the given curve. In each quadrilateral the diagonals are distinguished: one diagonal is dotted and the other is dashed.

Now imagine that the leftmost quadrilateral is continuously deformed to the rightmost quadrilateral by moving the vertices along the curve. The leftmost quadrilateral has dashed diagonal longer than the dotted diagonal. The rightmost quadrilateral has dashed diagonal shorter than the dotted diagonal. It is plausible therefore that at some intermediate position the dotted and the dashed diagonal have equal length (Figure 11.6). If we could also arrange for these two equal diagonals to meet at their midpoints, then the quadrilateral would have to be a rectangle.

What we have just described is a very simple form of what is called the

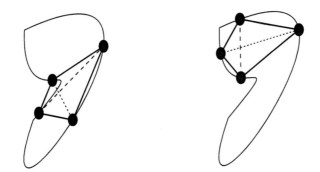

Figure 11.5: One diagonal longer than the other.

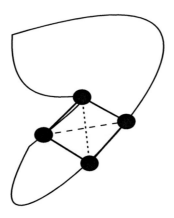

Figure 11.6: Diagonals of equal length.

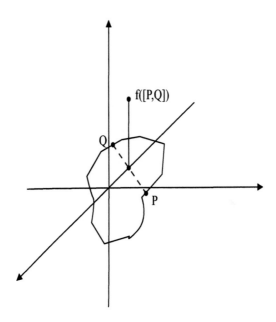

Figure 11.7: The function f.

"continuity method." In fact this argument alone is not sufficient to locate the desired rectangle. We need a more sophisticated continuity method that will keep track both of the lengths of the diagonals and of their centers at the same time. We caution the reader that a couple of sophisticated ideas are used along the way. Do not expect to completely understand the reasoning the first time through. But take it as an invitation to some new geometrical ideas.

Consider pairs of elements of the curve γ, but identify the pair (P, Q) with the pair (Q, P). In other words, consider *unordered* pairs of elements of γ (we will come back to this idea momentarily, and clarify what geometrical object this actually is). Denote the set of these unordered pairs with the letter S. Denote an element of S, that is an unordered pair of points on γ, by $\{P, Q\}$.

Now think of the curve γ as lying in the x-y plane in 3-dimensional space. For each $\{P, Q\} \in S$ we find the midpoint of the segment $\overline{PQ}$ and then let $f(\{P, Q\})$ be the point in space that lies a distance $|P - Q|$ above that midpoint. See Figure 11.7. This f is a continuous function from S into 3-dimensional Euclidean space.

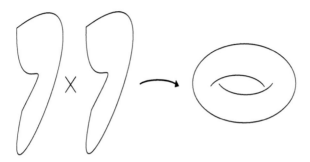

Figure 11.8: $\gamma \times \gamma$ is geometrically a torus.

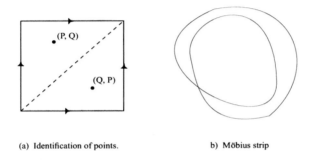

(a) Identification of points.   b) Möbius strip

Figure 11.9: The torus with identifications.

Now, what geometrical object is S? The set of ordered pairs of elements of γ corresponds in a natural way to a torus—see Figure 11.8.

However we are making the situation more complex by identifying (P, Q) with (Q, P). The markings in Figure 11.9(a) suggest how this is done. The resulting geometric object is what is called a "non-orientable" surface—in fact it is a Möbius strip (see Figure 11.9(b)).

But the function f is a continuous function of S into three-dimensional space. If f were one-to-one, then its image would be a realization of S as a subset of three-dimensional space. This is not yet a contradiction, because the torus with the additional identifications that we have specified is equivalent to the Möbius strip. But we now obtain a contradiction as follows. The bounding edge of the embedded surface is a simple closed curve. Paste a topological "disc" to it to make the surface closed. The result is a "cross cap"—in other words, it is a realization of the projective plane as an embedded surface in three-dimensional space. This is known to be impossible. [You may wish to get some help with this idea.] Therefore f cannot be one-to-one.

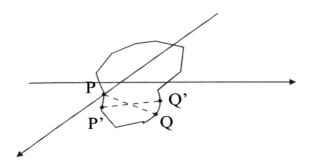

Figure 11.10: Diagonals of equal length.

What does this mean?

It means that there are two (unordered) pairs $\{P, Q\}$ and $\{P', Q'\}$ that have the same image under f. This means that the segment $\overline{PQ}$ and the segment $\overline{P'Q'}$ have the same midpoint; moreover, it means that $\overline{PQ}$ and $\overline{P'Q'}$ have the same *length*, for the heights of the points $f(\{P, Q\})$ and $f(\{P', Q'\})$ above the x-y plane are the same.

But now look at Figure 11.10. The segments $\overline{PQ}$ and $\overline{P'Q'}$ having the same length and the same midpoint implies that they are the diagonals of a rectangle. In other words, the points P, Q, P', Q' are four points on the curve γ that are the corners of a rectangle.   □

As of this writing, it is an open problem to determine whether any curve γ as in this last problem has on it four points that are the corners of a square.

Now let us think about collinearity. Let $P_1, P_2, P_3, \ldots, P_k$ be points in the plane—just finitely many of them—that are not all collinear. Show that there is a line in the plane that passes through just two of these points.

Just to illustrate the ideas, look at Figure 11.11. The first part exhibits three points, not all collinear. It also shows a line that passes through just two of them. The second part of Figure 11.11 shows four points, not all collinear. It also shows a line that passes through just two of them.

If we are given a large, but finite, collection of points—configured in a fairly arbitrary manner—then how might we find two of them through which

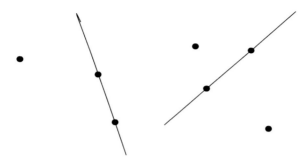

Figure 11.11: Noncollinear points and a line passing through just two of them.

a line passes that contains none of the other points P_j? We solve this problem by using one of the most powerful techniques in all of mathematics: that is, we solve an extremal problem.

Consider the set T of all ordered pairs (ℓ, P_m) consisting of a line that passes through *at least* two of the P_j's and a point P_m *not on that line* (such a point P_m must exist since the points are assumed to not all be collinear). Define a function

$$f : T \to \mathbb{R}$$

by

$$f((\ell, P_m)) = \text{distance of } \ell \text{ to } P_m.$$

Notice that f always takes positive values. Also the domain of f is a finite set (since there are only finitely many lines ℓ and finitely many points P_m). Therefore there will be a pair $(\widetilde{\ell}, \widetilde{P_m})$ at which f takes a minimum value. We claim that this line $\widetilde{\ell}$ is the one we seek.

Examine Figure 11.12. It shows one possible configuration of the line $\widetilde{\ell}$ and the point $\widetilde{P_m}$. Recall that, by definition, $\widetilde{\ell}$ is a line that contains (at least) two of the points P_j. One possible configuration of two of those points is shown in Figure 11.12. We claim that $\widetilde{\ell}$ could not contain a third point P_r.

If P_r were situated as in Figure 11.13, then the line ℓ' through $\widetilde{P_m}$ and P_r would be closer to P_t than $\widetilde{P_m}$ is to $\widetilde{\ell}$.

We have just argued, in detail, that if the minimal pair $(\widetilde{\ell}, \widetilde{P_m})$ is configured as in Figure 11.12, then $\widetilde{\ell}$ can have only two of the P_j's on it. We

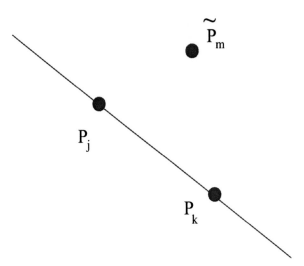

Figure 11.12: An extremal line and point.

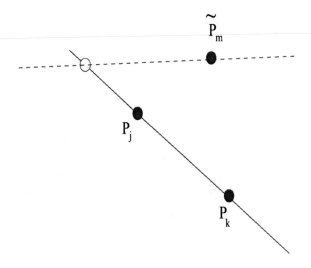

Figure 11.13: Impossibility of a third point.

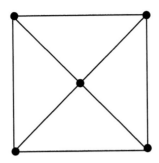

Figure 11.14: Impossibility of an Euler circuit.

invite you to consider other possible configurations of $(\widetilde{\ell}, \widetilde{P}_m)$ and argue that no third point P_r could be on the line $\widetilde{\ell}$—otherwise there would be a contradiction to the minimality of the function f on the pair $(\widetilde{\ell}, \widetilde{P}_m)$.

In conclusion, the line $\widetilde{\ell}$ taken from the pair $(\widetilde{\ell}, \widetilde{P}_m)$ that minimizes f is a line passing through just two of the points $P_1, P_2, P_3, \ldots, P_k$.

As a challenge, show that the assertion of the last problem can be false if the given collection of points has infinitely many elements. **Hint:** Think about the integer lattice.

As a second challenge, find a solution of the last problem that does not rely on an extremal problem, but instead breaks the issue up into cases.

Exercise: The graph shown in Figure 11.14 cannot be traversed in such a way that each side is traveled once and only once. Explain why.

11.5 Scheduling Sporting Events

The creation of the schedule for a sport can be quite complicated. For example, there are 30 professional baseball teams in the United States, and each plays 162 games per year. Each team wants to play a certain number of home games and a certain number of away games. Each team wants to play each other team a certain number of times. Team A typically wants to play Team B in a group of three games. And there are many other constraints as well.

The travel costs for Team A to travel to the stadium of Team B are considerable (for not only the team members but all the support staff must be transported). So there is certainly interest in finding ways to minimize these travel costs.

In recent years, mathematicians have gotten involved in the problem. In fact a mathematical company called Sports Scheduling Group has created the major league schedule for professional baseball in six of the last seven seasons. Another mathematical company has applied to create the schedule for the Nippon Professional Baseball league in Japan. It should be noted that the Sports Scheduling Group has saved major league baseball about 25% of its travel costs.

Our purpose in this section is to describe how to reduce this sports-team-scheduling problem to a purely mathematical problem in graph theory. We shall not describe the complete solution to the problem, as that is too complicated. But we can at least give you a flavor of how the problem is attacked.

Because it is more manageable than American major league baseball, we shall consider scheduling for Nippon Professional Baseball (NPB). NPB is split into the six-team Pacific League (PL) and the six-team Central League (CL). Each team plays 144 games during the regular season. Of these, 120 of the games are *intra-league*, meaning against teams from their own league. And 24 games are *inter-league*, meaning against teams from the other league.

The PL teams are Fukuoka, Orix, Saitama, Chiba, Tohoku, and Hokkaido. It will be convenient to refer to these as p_1, p_2, p_3, ..., p_6. The CL teams are Hiroshima, Hanshin, Chunichi, Yokohama, Yomiuri, and Yakult. We refer to these as c_1, c_2, c_3, ..., c_6.

To be more specific, we note that each NPB team plays 12 home games and 12 away games against each of the other 5 teams in its league (giving $24 \times 5 = 120$ games) and also 2 homes games and 2 away games against all 6 teams in the other league (giving $4 \times 6 = 24$ games). And that gives the total of $120 + 24 = 144$ games in a season.

As in American major league baseball, virtually all NPB games occur in *sets* of 3 games, played in a row. So we take it, for intra-league play, that each team plays 40 sets of 3 games. For inter-league play, each team will play 12 sets of 2 games.

For each team, we define a *trip* to be any pair of consecutive sets not occurring in the same venue. That is, a trip is any situation where a team has to travel from one venue to another to play its next set of games. In Table 11.1, we list the total distance traveled by all teams under the mathematically optimized schedule, and compare that distance to the actual distance traveled by the teams in the year 2010. There are also statistics for the number of total trips taken by the teams. Notice that the mathematically optimized schedule saves nearly 70,000 kilometers.

Table 11.1. The distance-optimal NPB schedule versus the actual 2010 regular-season schedule.

	Dist. (2010)	Dist. (Optim.)	Reduct. in Dist.	Trips (2010)	Trips (Optim.)	Reduct. in Trips
Intra-Lg. (PL)	153, 940	114, 169	25.8%	208	169	18.8%
Intra-Lg. (CL)	79, 067	57, 836	26.8%	199	170	14.6%
Inter-Lg.	51, 134	42, 950	16.0%	108	101	6.5%
Total	284, 141	214, 955	24.3%	515	440	14.6%

The discussion here will treat an analysis of the problem for intra-league play. That for inter-league play is similar, and we do not treat it. The ideas presented here are taken from [HOK].

Let us examine the intra-league schedule presented in Table 11.2. It shows $n = 6$ teams (as in the PL) each playing $k = 4$ *blocks* of 10 sets. Recall that a set is a triplet of games. Each block consists of 2 *rounds*, and each round has $n - 1 = 5$ sets, and each set has 3 games. In summary, each team plays a total of $k \cdot 2 \cdot (n - 1) = 40$ sets, for a total of 120 games.

If n and k are positive integers (with n and k as in the preceding paragraph), let D be the matrix whose entries are the distances between the home stadiums. In particular, the $D_{i,j}$ (*i*th row and *j*th column) entry of the matrix is the distance between the stadiums of teams i and j. Obviously $D_{i,j} = D_{j,i}$ for all i, j and $D_{i,i} = 0$ for all i. We call D the *distance matrix*.

For any pair n, k of positive integers and corresponding distance matrix D, what we wish to achieve is an intra-league tournament schedule that minimizes the total distance traveled by all n teams, subject to the following conditions (and we note that it is these various constraints that make the problem interesting):

(a) **The compactness condition:** The tournament lasts $k \cdot 2 \cdot (n-1)$ sets, which is $2k$ rounds, where each team has one set scheduled in each time slot.

(b) **The each round condition:** Each pair of teams must play exactly once per round, with their matches in rounds $2t-1$ and $2t$ taking place in different venues (for all $1 \le t \le k$).

(c) **The at-most-three condition:** No team may have a home stand or a road trip lasting more than three sets.

(d) **The no-repeat condition:** A team cannot play against the same opponent in two consecutive sets.

(e) **The diff-two condition:** Let $H_{i,s}$ and $R_{i,s}$ be the number of home and away sets played by team i within the first s sets. Then $|H_{i,s} - R_{i,s}| \le 2$ for all i, s with $1 \le i \le n$ and $1 \le s \le k \cdot 2 \cdot (n-1)$.

It is straightforward to verify that Table 11.2 satisfies these five conditions. Note, in Table 11.2, that an underlined set is one that is played at home. So, for instance, in the first round for team p_1, there are home games with p_2 and p_6 and away games with p_5, p_3, p_4. And in the second round for team p_1, there are home games with p_3, p_4, and p_5 and away games with p_6, p_2. Thus we see that the travel schedule for p_1 in the first block (consisting of the two rounds just described) is $D_{p_1,p_5} + D_{p_5,p_3} + D_{p_3,p_4} + D_{p_4,p_1} + D_{p_1,p_6} + D_{p_6,p_2} + D_{p_2,p_1}$.

Table 11.2. A 40-set (120-game) intra-league schedule for the NPB Pacific League.

Tm.	R1	R2	R3	R4	R5	R6	R7	R8
P1	$\underline{P2}\,P5\,P3\,P4\,\underline{P6}$	$\underline{P3}\,\underline{P4}\,P6\,P2\,\underline{P5}$	$\underline{P2}\,P4\,P6\,P5\,P3$	$\underline{P6}\,\underline{P5}\,P3\,P2\,P4$	$P2\,\underline{P3}\,\underline{P4}\,P5\,P6$	$P4\,\underline{P5}\,\underline{P6}\,\underline{P2}\,P3$	$P6\,P2\,\underline{P3}\,\underline{P5}\,\underline{P4}$	$P3\,P5\,P4\,\underline{P6}\,\underline{P2}$
P2	$P1\,\underline{P3}\,\underline{P6}\,\underline{P5}\,P4$	$P6\,P5\,\underline{P4}\,\underline{P1}\,P3$	$P1\,\underline{P3}\,\underline{P4}\,P6\,P5$	$P4\,\underline{P6}\,\underline{P5}\,\underline{P1}\,P3$	$\underline{P1}\,P4\,P5\,P6\,P3$	$\underline{P5}\,\underline{P6}\,\underline{P3}\,P1\,P4$	$P3\,\underline{P1}\,\underline{P5}\,\underline{P4}\,P6$	$P5\,P4\,\underline{P6}\,\underline{P3}\,P1$
P3	$P4\,P2\,\underline{P1}\,\underline{P6}\,\underline{P5}$	$P1\,P6\,P5\,\underline{P4}\,\underline{P2}$	$\underline{P6}\,P2\,P5\,\underline{P4}\,\underline{P1}$	$P5\,P4\,P1\,P6\,\underline{P2}$	$\underline{P5}\,P1\,P6\,P4\,\underline{P2}$	$\underline{P6}\,\underline{P4}\,P2\,P5\,\underline{P1}$	$\underline{P2}\,P4\,P1\,\underline{P6}\,\underline{P5}$	$\underline{P1}\,P6\,P5\,P2\,\underline{P4}$
P4	$\underline{P3}\,P6\,P5\,\underline{P1}\,\underline{P2}$	$\underline{P5}\,P1\,P2\,P3\,\underline{P6}$	$\underline{P5}\,P1\,P2\,P3\,\underline{P6}$	$\underline{P2}\,\underline{P3}\,P6\,P5\,\underline{P1}$	$\underline{P6}\,P2\,P1\,\underline{P3}\,\underline{P5}$	$\underline{P1}\,P3\,P5\,P6\,\underline{P2}$	$P5\,\underline{P3}\,\underline{P6}\,P2\,P1$	$\underline{P6}\,\underline{P2}\,\underline{P1}\,\underline{P5}\,P3$
P5	$\underline{P6}\,\underline{P1}\,\underline{P4}\,P2\,P3$	$P4\,\underline{P2}\,\underline{P3}\,\underline{P6}\,P1$	$P4\,\underline{P6}\,\underline{P3}\,\underline{P1}\,P2$	$P3\,P1\,P2\,\underline{P4}\,\underline{P6}$	$P3\,\underline{P6}\,\underline{P2}\,\underline{P1}\,P4$	$P2\,P1\,\underline{P4}\,\underline{P3}\,P6$	$\underline{P4}\,\underline{P6}\,P2\,P1\,P3$	$\underline{P2}\,\underline{P1}\,P3\,P4\,P6$
P6	$P5\,\underline{P4}\,P2\,P3\,P1$	$\underline{P2}\,\underline{P3}\,\underline{P1}\,P5\,P4$	$P3\,\underline{P5}\,\underline{P1}\,\underline{P2}\,P4$	$P1\,P2\,\underline{P4}\,\underline{P3}\,P5$	$P4\,P5\,\underline{P3}\,\underline{P2}\,\underline{P1}$	$P3\,P2\,P1\,\underline{P4}\,\underline{P5}$	$P1\,\underline{P5}\,\underline{P4}\,\underline{P3}\,P2$	$\underline{P4}\,\underline{P3}\,P2\,P1\,\underline{P5}$

Recall that a block consists of two rounds, and each round consists of $n-1 = 5$ teams that will be played (in a set of $n/2 = 3$). So we may think of each column of a block as a set consisting of $n/2$ different *matches*; each match should specify the home team and the away team. Refer to Section 10.2 for the ideas about counting that we are about to use.

For any column in the block, there are $\binom{n}{n/2}$ ways to select the home team. Also, there are $\binom{n}{n/2} \cdot (n/2)!$ ways to specify the matches for any column, since there are $(n/2)!$ ways to map any choice of $n/2$ home teams to the unselected $n/2$ away teams to decide the set of $n/2$ matches. As a result, there are $m = \binom{n}{n/2}^2 \cdot (n/2)!$ different ways that we can specify matches of the first column and the home teams of the second column. For $n = 6$ (as in the NPB), we see that

$$m = \binom{6}{3}^2 \cdot 3! = 2400 \,.$$

There are m ways that the first two columns of a block can be chosen as we have described, with the first column specifying matches and the second column specifying home teams. There are also m different ways to specify the last two columns of a block.

The problem of creating a schedule for the league consists, roughly speaking, of creating a "path" from those first two columns to those last two columns. Since there are so many choices (thousands) in each case, you can see that this is a complicated problem.

The problem is completely solved in the article [HOK], and its authors have gone to work now for the NPB. They are saving money for the league, and also contributing to the saving of fuel and the greening of the environment. A happy ending for all.

Exercise: A chess tournament is to have 12 players. Each player is to play each other player 3 times. Four games can go on simultaneously. There can be 2 sessions per day (4 simultaneous games in each session). We want to schedule this tournament so that it is over in as few days as possible. Construct a schedule for this tournament.

Chapter 12

Advanced Ideas

12.1 Searching on Google

12.1.1 The Mathematics of a Google Search

Roughly 94 million Americans—about one third the population of the United States—use the Internet each day. Although there are many Internet search engines available, there is no question that the most popular one is Google.

The thing that sets Google apart from other search engines is its use of PageRank to rank pages found in a Web search. It is generally felt that PageRank is very effective, and gives more attractive results than the competitors. And PageRank is based entirely on mathematics—particularly on matrix theory.

Sergey Brin and Larry Page, formerly computer science graduate students at Stanford University, are the founders of Google. The idea for PageRank is largely due to Larry Page. Most search engines use Web page content as a device for ordering the pages found in a search. By contrast, PageRank uses the link structure of the Web to determine the importance of a Web page. One advantage of the PageRank technology is that its orderings cannot be manipulated by information concealed in Web pages.

PageRank is widely recognized as a superior method of ranking Web pages. Both the business and academic communities hold PageRank in high regard. Maximizing the PageRank score of a Web page has become an important

269

component of company marketing strategies. PageRank is not only connected (at least theoretically) to matrix theory, but also to numerical analysis, information retrieval, and graph theory. Research to understand and to optimize PageRank is ongoing.

Let us summarize the PageRank philosophy here. A typical Web surfer randomly chooses a Web page to view from the listing of available Web pages. Then he/she randomly selects a link from the Web page to go to another Web page. The surfer continues the process of iteratively choosing links from successive Web pages until finally deciding to move to another Web page by some means *other* than choosing a link. This last choice does *not* depend on the previously visited Web pages, and the Web surfer never tires of visiting Web pages. In summary, the PageRank score of a Web page represents the probability that a random Web surfer chooses to view the Web page.

It is a pleasure to acknowledge here the well-written source [WIL], which provided both background and key examples for our discussion.

12.1.2   The Directed Web Graph

In order to come up with a mathematical model for the activity of the typical Web surfer, the PageRank algorithm depicts the link structure of the Web as a directed graph. Each Web page is a node of the graph. The link from one Web page to another Web page is an arrow: in other words, these are edges that exhibit direction of movement. In practice this graph could be very large—involving billions of pages. But we can apply PageRank to a directed graph of any size. For illustrative purposes here we restrict attention to a quite small graph. Examine Figure 12.1.

12.1.3   Passage to the Web HyperLink Matrix

The key idea now is to convert the directed graph from the last subsection to a matrix. For the graph exhibited in Figure 12.1, this will be a 4×4 matrix that we call H. This is the *Web hyperlink matrix*. The main point is that, if Web page j has ℓ_j links to other Web pages and if Web page j links to Web page k, then the element in row j and column k of the matrix H is $H_{jk} = 1/\ell_j$. Otherwise $H_{jk} = 0$. Thus H_{jk} represents the likelihood that a

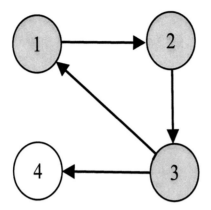

Figure 12.1: A directed graph with four nodes.

random surfer will select a link from Web page j to Web page k. For the directed graph in Figure 12.1, we thus see that

$$H = \begin{pmatrix} 0 & 1 & 0 & 0 \\ 0 & 0 & 1 & 0 \\ 1/2 & 0 & 0 & 1/2 \\ 0 & 0 & 0 & 0 \end{pmatrix}.$$

Observe here that node 4 is a *dangling node* because it does not link to other nodes. As a result, all entries in row 4 of H are 0. In concrete terms, this means that the likelihood that a surfer will move from node 4 to any other node in the directed graph is 0. It turns out that the majority of Web pages are postscript files and image files and are therefore dangling nodes. It follows that a great many of the rows in the Web hyperlink matrix are all zeros. When a Web surfer lands on a dangling node Web page, he/she can either stop surfing (perish the thought!) or jump to a different Web page by entering a Web address in the browser address line. Since H does not report on the possibility of moving from a dangling node Web page to another Web page, the long-term behavior of Web surfers cannot be completely determined from H alone.

12.1.4 A Fix for Dangling Nodes

There are several possible options for modeling the behavior of a Web surfer after he/she lands on a dangling node. Google does not reveal which option it employs.   One possibility is to replace each dangling row (that is, row of all zeros) of H with a particular *probability distribution vector* w.   This vector w is a vector with nonnegative elements that sum to 1.   Let d be a column vector that has a 1 in the jth place if $\ell_j = 0$ and a 0 in the jth place otherwise.   The way to write our new idea in vector-matrix language is

$$S = H + dw.$$

Notice that we have written dw, which is the product of a 4×1 column vector (on the left) times a 1×4 row vector (on the right).   The result is a 4×4 matrix.

More explicitly, with

$$d = \begin{pmatrix} d_1 \\ d_2 \\ d_3 \\ d_4 \end{pmatrix}$$

and

$$w = (w_1, w_2, w_3, w_4),$$

we have

$$dw = \begin{pmatrix} d_1 \\ d_2 \\ d_3 \\ d_4 \end{pmatrix} \cdot (w_1, w_2, w_3, w_4).$$

Now the way that we multiply two matrices of this kind is as follows:

- We multiply the first element of d times the first element of w and place the answer in the $(1, 1)$ position in the product (the answer) matrix.

- We multiply the first element of d times the second element of w and place the answer in the $(1, 2)$ position in the product (the answer) matrix.

- We multiply the first element of d times the third element of w and place the answer in the $(1, 3)$ position in the product (the answer) matrix.

- We multiply the first element of d times the fourth element of w and place the answer in the $(1, 4)$ position in the product (the answer) matrix.

That fills out the first row of the product matrix.

Next we do the following:

- We multiply the second element of d times the first element of w and place the answer in the $(2, 1)$ position in the product (the answer) matrix.

- We multiply the second element of d times the second element of w and place the answer in the $(2, 2)$ position in the product (the answer) matrix.

- We multiply the second element of d times the third element of w and place the answer in the $(2, 3)$ position in the product (the answer) matrix.

- We multiply the second element of d times the fourth element of w and place the answer in the $(2, 4)$ position in the product (the answer) matrix.

That fills out the second row of the product matrix.

Next we do the following:

- We multiply the third element of d times the first element of w and place the answer in the $(3, 1)$ position in the product (the answer) matrix.

- We multiply the third element of d times the second element of w and place the answer in the $(3, 2)$ position in the product (the answer) matrix.

- We multiply the third element of d times the third element of w and place the answer in the $(3, 3)$ position in the product (the answer) matrix.

- We multiply the third element of d times the fourth element of w and place the answer in the $(3, 4)$ position in the product (the answer) matrix.

That fills out the third row of the product matrix.

Next we do the following:

- We multiply the fourth element of d times the first element of w and place the answer in the $(4, 1)$ position in the product (the answer) matrix.

- We multiply the fourth element of d times the second element of w and place the answer in the $(4, 2)$ position in the product (the answer) matrix.

- We multiply the fourth element of d times the third element of w and place the answer in the $(4, 3)$ position in the product (the answer) matrix.

- We multiply the fourth element of d times the fourth element of w and place the answer in the $(4, 4)$ position in the product (the answer) matrix.

That fills out the fourth row of the product matrix.

As a result, the product matrix is

$$
dw = \begin{pmatrix}
d_1 w_1 & d_1 w_2 & d_1 w_3 & d_1 w_4 \\
d_2 w_1 & d_2 w_2 & d_2 w_3 & d_2 w_4 \\
d_3 w_1 & d_3 w_2 & d_3 w_3 & d_3 w_4 \\
d_4 w_1 & d_4 w_2 & d_4 w_3 & d_4 w_4
\end{pmatrix}
$$

As you can see, this is a 4×4 matrix as we asserted above.

The most common choice for the row vector w is

$$
w = (1/4, 1/4, 1/4, 1/4).
$$

This in effect adds an artificial link from each dangling node to every other Web page. With this choice of w, the changes in the directed graph are shown in Figure 12.2.

Of course the four dotted arrows in Figure 12.2 come from the row of $1/4$s that we have added to the matrix. The new matrix $S = H + dw$ is in fact

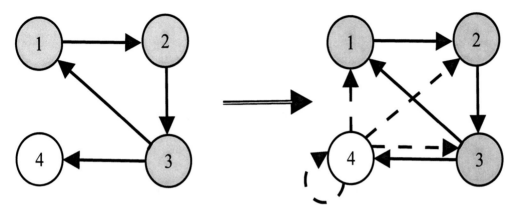

Figure 12.2: Modified directed graph with dangling nodes fixed.

$$S = \begin{pmatrix} 0 & 1 & 0 & 0 \\ 0 & 0 & 1 & 0 \\ 1/2 & 0 & 0 & 1/2 \\ 0 & 0 & 0 & 0 \end{pmatrix} + \begin{pmatrix} 0 \\ 0 \\ 0 \\ 1 \end{pmatrix} \begin{pmatrix} 1/4 & 1/4 & 1/4 & 1/4 \end{pmatrix}$$

$$= \begin{pmatrix} 0 & 1 & 0 & 0 \\ 0 & 0 & 1 & 0 \\ 1/2 & 0 & 0 & 1/2 \\ 1/4 & 1/4 & 1/4 & 1/4 \end{pmatrix}.$$

There are several options for dealing with dangling nodes. No matter which option is chosen, Google will create a new matrix S that models the tendency of a Web surfer to leave a dangling node. Nonetheless, our model is not yet complete.

For, even when Web pages have links to other Web pages, a surfer might grow tired of clicking on links; he/she might decide to move to a different Web page in some other way. Notice, in the right-hand picture in Figure 12.2, there is no directed edge from node 2 to node 1. On the Web, however, a surfer can move directly from node 2 to node 1 by entering the Web address for node 1 in the address line of the Web browser. The matrix S does not acknowledge this possibility.

12.1.5   The Ultimate Google Matrix

In order to model the overall behavior of a typical Web surfer, we must make the following final modification. Define now the matrix

$$G = \alpha S + (1 - \alpha)\mathbf{1}v \,,$$

where $\mathbf{1}$ is the column matrix of all 1s and v is a row probability distribution vector called the *personalization vector*. We think of α (the damping factor) as a probability, so it is a number between 0 and 1. Specifically, the Web surfer moves to a different Web page by a means *other* than a link with probability $(1 - \alpha)$. Experiments conducted by Brin, Page, and others suggest that a good choice for α is 0.85 and a good choice for v is $v = (1/4, 1/4, 1/4, 1/4)$. In most research work on the PageRank algorithm, values for α that appear range from 0.85 to 0.99.

Notice that each element G_{ij} of G lies between 0 and 1. And the sum of the elements in each row is 1. So we call the Google matrix GA a *row stochastic matrix*. The equation

$$\pi G = \pi \,,$$

where π is a row probability distribution vector, will have just one solution. So call π the corresponding *dominant left eigenvector* of G. It is π that is the key to PageRank. Specifically,

> the jth entry of π is the PageRank score for Web page j, and π is called the PageRank vector.

We want to illustrate the use of the PageRank vector, and how to interpret it. To do so, we now present Table 12.1 showing different PageRank vectors that arise from different damping factors and different personalization vectors.

Table 12.1 Models of Web surfer behavior for the directed graph in Figure 12.2.

Damp. Factor (α)	Personaliz. Vector (v)	Google Matrix (G)	Pgrnk Vector $(\approx \pi)$	Ord. of Nodes (1 Hghst.)
Mdl. 1 0.85	$(\frac{1}{4}, \frac{1}{4}, \frac{1}{4}, \frac{1}{4})$	$\begin{pmatrix} \frac{3}{80} & \frac{71}{80} & \frac{3}{80} & \frac{3}{80} \\ \frac{3}{80} & \frac{3}{80} & \frac{71}{80} & \frac{3}{80} \\ \frac{37}{80} & \frac{3}{80} & \frac{3}{80} & \frac{37}{80} \\ \frac{1}{4} & \frac{1}{4} & \frac{1}{4} & \frac{1}{4} \end{pmatrix}$	$(0.21, 0.26, 0.31, 0.21)$	(3 2 1 3)
Mdl. 2 0.85	$(1, 0, 0, 0)$	$\begin{pmatrix} \frac{3}{20} & \frac{17}{20} & 0 & 0 \\ \frac{3}{20} & 0 & \frac{17}{20} & 0 \\ \frac{23}{40} & 0 & 0 & \frac{17}{40} \\ \frac{29}{80} & \frac{17}{80} & \frac{17}{80} & \frac{17}{80} \end{pmatrix}$	$(0.30, 0.28, 0.27, 0.15)$	(1 2 3 4)
Mdl. 3 0.95	$(\frac{1}{4}, \frac{1}{4}, \frac{1}{4}, \frac{1}{4})$	$\begin{pmatrix} \frac{1}{80} & \frac{77}{80} & \frac{1}{80} & \frac{1}{80} \\ \frac{1}{80} & \frac{1}{80} & \frac{77}{80} & \frac{1}{80} \\ \frac{39}{80} & \frac{1}{80} & \frac{1}{80} & \frac{39}{80} \\ \frac{1}{4} & \frac{1}{4} & \frac{1}{4} & \frac{1}{4} \end{pmatrix}$	$(0.21, 0.26, 0.31, 0.21)$	(3 2 1 3)
Mdl. 4 0.95	$(1, 0, 0, 0)$	$\begin{pmatrix} \frac{1}{20} & \frac{19}{20} & 0 & 0 \\ \frac{1}{20} & 0 & \frac{19}{20} & 0 \\ \frac{21}{40} & 0 & 0 & \frac{19}{40} \\ \frac{23}{80} & \frac{19}{80} & \frac{19}{80} & \frac{19}{80} \end{pmatrix}$	$(0.24, 0.27, 0.30, 0.19)$	(3 2 1 4)

Let us do some calculations to explain exactly how each entry in the first row of this table was obtained.

First we note that α is a choice we make. Values of $0.85 = 17/20$ and higher for α are quite common in practice. The choice of v is also up to the user. The choice $v = (1/4, 1/4, 1/4, 1/4)$ is an intuitively appealing one and commonly used. But the choice $v = (1, 0, 0, 0)$ also makes good sense.

Once these two choices are made, then all the other entries are calculated.

We know from the discussion above that

$$
S = \begin{pmatrix}
0 & 1 & 0 & 0 \\
0 & 0 & 1 & 0 \\
1/2 & 0 & 0 & 1/2 \\
1/4 & 1/4 & 1/4 & 1/4
\end{pmatrix}.
$$

Furthermore, using the values of α and v specified for Model 1, we have

$$
G = \alpha S + (1 - \alpha) \gg v
$$

$$
= \frac{17}{20} \begin{pmatrix}
0 & 1 & 0 & 0 \\
0 & 0 & 1 & 0 \\
1/2 & 0 & 0 & 1/2 \\
1/4 & 1/4 & 1/4 & 1/4
\end{pmatrix} + \frac{3}{20} \begin{pmatrix}
1 \\ 1 \\ 1 \\ 1
\end{pmatrix} (1/4, 1/4, 1/4, 1/4)
$$

$$
= \begin{pmatrix}
0 & 17/20 & 0 & 0 \\
0 & 0 & 17/20 & 0 \\
17/40 & 0 & 0 & 17/40 \\
17/80 & 17/80 & 17/80 & 17/80
\end{pmatrix} + \begin{pmatrix}
3/80 & 3/80 & 3/80 & 3/80 \\
3/80 & 3/80 & 3/80 & 3/80 \\
3/80 & 3/80 & 3/80 & 3/80 \\
3/80 & 3/80 & 3/80 & 3/80
\end{pmatrix}
$$

$$
= \begin{pmatrix}
3/80 & 71/80 & 3/80 & 3/80 \\
3/80 & 3/80 & 71/80 & 3/80 \\
37/80 & 3/80 & 3/80 & 37/80 \\
1/4 & 1/4 & 1/4 & 1/4
\end{pmatrix}.
$$

Notice that, as predicted, the entries of this matrix are all positive and the entries in each row add up to one.

We shall not go through the procedure of actually find the eigenvector of this matrix, but we shall verify that (up to a roundoff error) the vector $\pi = (0.21, 0.26, 0.3, 0.21)$ is that eigenvector. Our job then is to calculate the product

$$
\pi G = (0.21, 0.26, 0.3, 0.21) \begin{pmatrix}
3/80 & 71/80 & 3/80 & 3/80 \\
3/80 & 3/80 & 71/80 & 3/80 \\
37/80 & 3/80 & 3/80 & 37/80 \\
1/4 & 1/4 & 1/4 & 1/4
\end{pmatrix}.
$$

This is just a tedious exercise in arithmetic, and we find that this product is given, up to two decimal places, by

$$
\pi G = (0.21, 0.26, 0.3, 0.21).
$$

So we indeed have that
$$\pi G = \pi \, .$$

We may conclude that $\pi = (0.21, 0.26, 0.3, 0.21)$ is the PageRank vector. Since 0.3 is the largest entry, we put a 1 in that position in the "ordering of nodes." Since 0.26 is the second largest entry, we put a 2 in that position in the "ordering of nodes." Since 0.21 is the third largest entry, we put a 3 in both those positions in the "ordering of nodes."

And this is how Google ranks Web pages. We see that, for us, the Web page 3 is the one that comes out on top. Looking back at Figure 12.2, it is not difficult to see why. For 3 is the node that is most fruitful in leading to other useful nodes.

Of course the example that we have worked out here is a "toy" version of the problem. It is unrealistic that there is a Web page with at most three links. More typical is millions or even billions of links. The resulting matrix will then be quite large. Fortunately, most of the entries in the matrix are zeros, so it is still feasible to perform the necessary computations.

Exercise: There are tricks in Google so that, if your Web page comes up in a Web search, then it will be near the top of the stack. Consider how to make this happen and describe how to do it.

12.2 A Needle Problem of Kakeya

We begin with a detailed discussion of a problem from the geometry of the plane.

The genesis of the elegant geometric problem just alluded to comes from a nineteenth-century Japanese scientist named S. Kakeya. He posed the following question (this is its classical formulation): Dip a sewing needle into a bottle of ink. Now place it on a piece of paper. Endeavor to move the needle on the piece of paper so as to (i) reverse the positions of the two ends, and (ii) leave behind an ink blot of smallest possible area.

This question baffled mathematicians for decades. It was finally solved by A. Besicovitch in 1928 (see [BES1], [BES2]).

Here are some preliminary thoughts about the question:

A closed disc in the plane.

A needle in the disc.

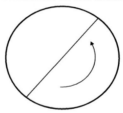

Rotating the needle.

Figure 12.3: Rotating a needle in a disc.

1. First suppose that the needle has length 2, and is of infinitesimal thickness. Let D be a closed disc in the plane of radius 1 (see Figure 12.3). Of course we may insert the needle inside D as shown in the figure.[1] Furthermore, the needle may be rotated so as to switch its ends, while keeping the needle inside the disc all the while: refer again to the figure. In the language of the Kakeya problem: If we let k be the greatest lower bound of areas of all ink blots that can be left behind by a needle as specified by Kakeya, then

$$k \leq \text{area of unit disc} = \pi \cdot 1^2 = \pi.$$

2. To improve on the estimate in **1**, consider Figure 12.4. Pin down one end of the needle, and swing the other end through an angle of $2\pi/3$

---

[1]Observe that, for clarity, we do not actually show the blot of ink—just the figure.

Figure 12.4: A reduced area estimate.

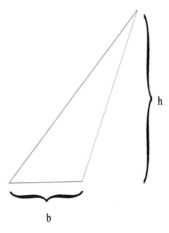

Figure 12.5: The sprouting process.

radians. This leaves behind an ink blot as shown. Now pin down the free end, and rotate the first end through an angle of $2\pi/3$ radians. See Figure 12.5. Now repeat the rotation process a third time. The resulting ink blot is shown in Figure 12.6. Clearly, by construction, this is a Kakeya blot. We invite the reader to verify that the *area* of this blot is $2\pi - 2\sqrt{3} \approx 2.81908\ldots$, a number that is strictly less than $\pi \approx 3.14159\ldots$—the area of the blot in the first example.

For many years, mathematicians thought that the blot described in this second example was in fact the solution to the Kakeya needle problem, that is, that this was the smallest blot possible.

Unlike some solutions to mathematics problems, which simply unleash an arsenal of weapons to confirm a predictable result, Besicovitch's solution to the Kakeya needle problem was a shock. For Besicovitch showed that, given

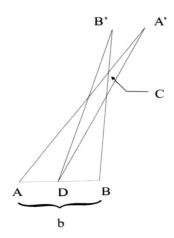

Figure 12.6: Details of the sprouting.

any $\epsilon > 0$ no matter how small, there is a set E in the plane such that **(i)** the area of E is less than ϵ, and **(ii)** the needle may have its ends switched, in the manner specified in the statement of the Kakeya needle problem, in such a way that the needle remains in the set E at all times. In other words, it is possible to move the needle in the manner specified by Kakeya but so that the ink blot left behind has area less than ϵ, no matter how small $\epsilon > 0$.

We now present a modern variant of the Besicovitch construction which may be found in 2[CUN]. It turns out to be a bit more flexible than Besicovitch's original solution to the problem, and also a bit easier to understand.

We begin with a triangle T of base b and height h, as shown in Figure 12.5. The basic step in our construction is to *sprout* T to two new triangles, of height $h' > h$. Here is what happens during the sprouting process; refer to Figure 12.6 as you read along. We extend AC to a point A' at height h' and likewise we extend BC to a point B' at height h'. We connect A' and B' to the midpoint D of AB, as shown. The two triangles $T'_A \equiv \triangle AA'D$ and $T'_B \equiv \triangle BB'D$ are called *sprouts* from height h to h'. See Figure 12.7. Of course we can later, if we wish, sprout the new triangles T'_A and T'_B to a new height $h'' > h'$. In fact we shall do this; more, we shall iterate our procedure some large positive number k times in order to obtain the desired estimates. The Kakeya set that we wish to construct will be, essentially, the union of all the sprouts. Refer to Figure 12.8.

Lemma 12.1 *With the triangle T and sprouts T'_A, T'_B specified as above,*

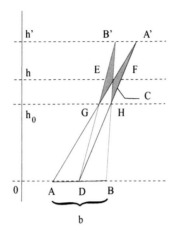

Figure 12.7: Area of the sprouts.

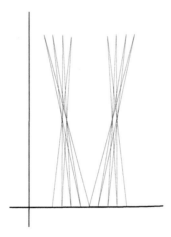

Figure 12.8: Movement of a needle through the sprouts.

the area of $T'_A \cup T'_B$ is

$$A_{T'_A \cup T'_B} = \frac{1}{2}bh + \frac{b(h'-h)^2}{2h'-h}.$$

Proof: Refer to Figure 12.7. Of course $\triangle B'DB$ is similar to $\triangle B'EC$, and $\triangle A'AD$ is similar to $\triangle A'CF$. Therefore

$$\frac{EC}{DB} = \frac{h'-h}{h'}$$

and

$$\frac{CF}{AD} = \frac{h'-h}{h'}.$$

As a result,

$$EC = \frac{h'-h}{h'} \cdot DB = \frac{h'-h}{h'} \cdot AD = CF.$$

Thus $EC = CF$; $AD = DB = b/2$. As a result,

$$EC = \frac{h'-h}{h'} \cdot \frac{b}{2}$$

and

$$CF = \frac{h'-h}{h'} \cdot \frac{b}{2}.$$

Now notice that, by proportionality, the height of the point G above the base of T is just the same as the height of the point H above the base of T; we let that common height be h_0. Also note that

$$\frac{h-h_0}{h_0} = \frac{EC}{AD},$$

which we see by comparing similar triangles $\triangle AGD$ and $\triangle CGE$. Thus

$$h - h_0 = \frac{EC}{AD} \cdot h_0$$

or

$$h = h_0 \left(\frac{EC}{AD} + 1 \right)$$

and

$$h_0 = \frac{h}{EC/AD + 1} = \frac{h}{EC/DB + 1} = \frac{h}{\frac{h'-h}{h'} + 1} = \frac{hh'}{2h'-h}.$$

The area of $\triangle ACB \cup \triangle GB'C \cup \triangle HCA'$ is thus

$$
\begin{aligned}
\frac{1}{2}bh &+ \left[\frac{1}{2}EC(h'-h) + \frac{1}{2}EC(h-h_0)\right] + \left[\frac{1}{2}CF(h'-h) + \frac{1}{2}CF(h-h_0)\right] \\
&= \frac{1}{2}bh + \frac{1}{2}EC(h'-h_0) + \frac{1}{2}CF(h'-h_0) \\
&= \frac{1}{2}bh + \frac{1}{2}\left(\frac{h'-h}{h'}\cdot\frac{b}{2}\right)\cdot(h'-h_0) + \frac{1}{2}\left(\frac{h'-h}{h'}\cdot\frac{b}{2}\right)\cdot(h'-h_0) \\
&= \frac{1}{2}bh + \left(\frac{h'-h}{h'}\cdot\frac{b}{2}\right)\left(h'-\frac{hh'}{2h'-h}\right) \\
&= \frac{1}{2}bh + \frac{b(h'-h)^2}{2h'-h}.
\end{aligned}
$$

This completes our calculation of the area of $A_{T'_A \cup T'_B}$. Observe in passing that each of the shaded triangles in the figure has area $b(h'-h)^2/[2(2h'-h)]$.
□

Now we wish to sprout the original triangle T times and then estimate the area of the union of all the sprouts. We take an initial equilateral triangle having base $b_0 = \epsilon$ and height $h_0 = \sqrt{3}\epsilon/2$. Thus the initial triangle T_0 is an equilateral triangle of side ϵ.

Now we define a sequence of heights

$$
\begin{aligned}
h_0 &= \frac{\sqrt{3}}{2}\epsilon \\
h_1 &= \frac{\sqrt{3}}{2}\left(1 + \frac{1}{2}\right)\epsilon \\
h_2 &= \frac{\sqrt{3}}{2}\left(1 + \frac{1}{2} + \frac{1}{3}\right)\epsilon \\
&\cdots \\
h_{j-1} &= \frac{\sqrt{3}}{2}\left(1 + \frac{1}{2} + \frac{1}{3} + \cdots + \frac{1}{j}\right)\epsilon.
\end{aligned}
$$

We sprout T_0 from height h_0 to height h_1 to obtain two new triangles. The new triangles are called T_{01}, T_{02}. Note that every sprout of every generation has base that is a sub-segment of AB. Thus the first-generation sprouts

T_{01} and T_{02} are certainly not disjoint. But the portions of T_{01} and T_{02} that protrude from $\triangle ABC$ have disjoint interiors (these are displayed as shaded triangles in Figure 12.7). We sprout each of T_{01} and T_{02} to the new height h_2 to obtain four new triangles T_{011}, T_{012}, T_{021}, and T_{022}. We continue the process, obtaining at the jth sprouting a total of 2^j new triangles of height h_j and base $2^{-j}\epsilon$. For a large positive integer k to be fixed later, the Besicovitch set E (spanning an angle of $\pi/3$) that we seek will be the union of the 2^k triangles generated at the kth sprouting.

Now let us estimate the area of E. At the jth sprouting, the increment of area (i.e., the area actually added at that stage) is composed of 2^j triangles of equal area. According to Lemma 8.1, and the comment at the end of the proof, each of these 2^j triangles has area

$$\epsilon^2 \cdot 2^{-(j-1)} \left(\frac{\sqrt{3}/2}{j+1}\right)^2 \cdot \frac{1}{2\left[2\frac{\sqrt{3}}{2}\left(1+\frac{1}{2}+\cdots+\frac{1}{j+1}\right) - \frac{\sqrt{3}}{2}\left(1+\frac{1}{2}+\cdots+\frac{1}{j}\right)\right]}.$$

After k sproutings, the region we will have created will have area

$$\left[\sqrt{3}/4 + \sum_{j=1}^{k} 2^j \cdot 2^{-(j-1)} \frac{[\sqrt{3}/2]/(j+1)^2}{2\left[2(1+\frac{1}{2}+\cdots+\frac{1}{j+1}) - (1+\frac{1}{2}+\cdots+\frac{1}{j})\right]}\right] \epsilon^2$$

$$\leq \left[\frac{\sqrt{3}}{4} + \sum_{j=1}^{k} \frac{\sqrt{3}}{2(j+1)^2}\right] \epsilon^2$$

$$\leq \left[\frac{\sqrt{3}}{2}\left(\frac{1}{2} + \left(\frac{\pi^2}{6} - 1\right)\right)\right] \epsilon^2$$

$$\leq 2\epsilon^2.$$

We see that the set E has area as small as we please, just so long as the triangle with which we begin has small area. Further, the angles of the sprouted triangles span a total angle of $\pi/3$. Notice (Figure 12.8) that a segment of length two can be moved—incrementally—through this angle of $\pi/3$ by utilizing each sprouted triangle in succession. By adjoining a total of six of these sprout regions E—each rotated through an angle of $\pi/3$ from the previous one—we obtain a region inside which our needle (segment) can be rotated through a full turn. [Precisely, the six sprouted triangles are joined so that all six of the initial triangles from which they grew now have the same center; this "star" now becomes the "nucleus" of the six collections of sprouts.]

Refer now to Figure 12.8. Place a needle at the edge of one of the sprouts. The needle can be moved through a small angle within that sprout, and then slid through the nucleus to a sprout on the opposite side. Then the needle can be moved through a small angle in this new sprout, and then slid through the nucleus to yet another new sprout on the opposite side. And so forth. And the area of the set we have created does not exceed $6 \cdot 2\epsilon^2$. Since $\epsilon > 0$ is arbitrarily small, the Kakeya needle problem is solved. Refer to [CUN] for the history of this fascinating problem and for all the details. That source also describes many variants of the problem. The source [GUZ] has an alternative, and very elegant, presentation of the Kakeya problem.

Exercise: Consider the analogue of the Kakeya problem in 3 dimensions. Can you suggest a solution and describe it?

12.3   Euclidean and Non-Euclidean Geometry

Perhaps the first systematic, rigorous treatment of a subject in mathematics was Euclid's version of geometry (~300 B.C.E.). While it is generally agreed that much of what Euclid recorded was due to scholars who came before Euclid, what is important about Euclid's contribution is the *manner in which he organized and recorded the subject.*

Euclid's geometry has five "undefinable" terms. These are *point, line,* the phrase "*lie on,*" the concept of *between,* and *congruent.* The notion of circle can in fact be defined in terms of these undefinables (i.e., the term "circle" itself is *not* an undefinable):

Definition 12.2 Let two points O and A be given. The set of all points P such that the segment OP is congruent to segment OA is called a *circle* with center O. Each of the segments OP is called a radius of the circle.

Euclid's geometry has five axioms (see [GRE]). They are as follows:

Figure 12.9: Euclid's second axiom.

P1 Through any pair of distinct points, there passes a line.

P2 For each segment $\overline{AB}$ and each segment $\overline{CD}$, there is a unique point E (on the line determined by A and B) such that B is between A and E and the segment CD is congruent to BE (Figure 12.9).

P3 For each point C and each point A distinct from C, there exists a circle with center C and radius CA.

P4 All right angles are congruent.

P5 For each line ℓ and each point P that does not lie on ℓ, there is a unique line m through P such that m is parallel to ℓ.

We have not been as careful here as would be necessary in a thorough, rigorous development of geometry. In particular, you should note that we have not provided all the necessary definitions. As an exercise, you may wish to study these axioms, determine which terms require definition, and provide those definitions. One important instance of our omissions occurs in Axiom **P5**: we did not first define "parallel." One working definition of two lines being parallel is that they do not intersect. Another working definition is that there exists a transversal that is perpendicular to both lines. You should also think about the definitions of "right angle" and "congruent segments" and "perpendicular." A thorough discussion of all these ideas appears in [GRE].

We know that the five axioms of Euclid's geometry are consistent because there is a *model* that satisfies these axioms. That model is the usual geometry that we learn in high school. It consists of the lines that we draw with a ruler and the circles that we draw with a compass. [We could digress at this point into a discussion of what it means to verify that said lines and circles satisfy the axioms, but instead we refer the reader to [GRE].]

For nearly 2000 years, scholars wrestled with the question of whether the first four axioms could be used to prove the fifth, the so-called parallel postulate. In other words, is **P5** *independent* of the other four axioms? In 1829, Nikolai Ivanovich Lobachewski (1793–1856) published the construction of a geometry that satisfies Euclid's first four axioms, but such that the parallel postulate fails. János Bolyai (1775–1856) constructed such a geometry at roughly the same time, but his ideas were published somewhat later. Carl Friedrich Gauss (1777–1855) claimed in a private communication to Bolyai's father that he had made these discoveries several years earlier, but had not published for a variety of enigmatic reasons. Gauss expressed his thoughts on the matter, in his letter to János Bolyai, as follows (see [GRE]):

> I am exceedingly sorry that I have failed to avail myself of our former great proximity to learn more of your work on the foundations of geometry; it surely would have saved me much useless effort and given me more peace, than one of my disposition can enjoy so long as so much is left to consider in a method of this kind. I have myself made much progress in this matter (though my other heterogeneous occupations have left me but little time for this purpose); though the course which you assure me you have reached as to the questioning of the truth of geometry. It is true that I have found much which many would accept as proof, but which in my estimation proves *nothing*, for instance, if it could be shown that a rectilinear triangle is possible, whose area is greater than that of any given surface, then I could rigorously establish the whole of geometry. Now most people, no doubt, would grant this as an axiom, but not I; it is conceivable that, however distant apart the vertices of the triangle might be chosen, this area might yet always be below a certain limit. I have found several other such theorems, but none of them satisfies me.

A delightful account of the history of these matters appears in [GRE].

In a moment we shall present a version of the non-Euclidean geometry of Bolyai and Lobachewski. First let us consider what it means. We have already acknowledged that there is a model for the five axioms of Euclid's geometry. In other words, in the classical rectilinear geometry of high school, Axioms **P1**, **P2**, **P3**, **P4**, and **P5** are true. So **P1**, **P2**, **P3**, **P4**, and **P5** are *consistent*. Now we are saying that there is a model for **P1**, **P2**, **P3**, **P4**, and $\sim$ **P5**. This says that **P1**, **P2**, **P3**, **P4**, and $\sim$ **P5** are consistent.[2] In other words, because of the standard Euclidean model, **P5** is consistent with the other four axioms; and because of the new model that we are about to present, it is the case that $\sim$ **P5** is consistent with the other four axioms. This means that neither **P5** nor $\sim$ **P5** could be derived from the other four axioms. In other words, the fifth axiom is independent of the first four.

Gauss claimed that he did not publish his construction of non-Euclidean geometry in part because nobody else would understand its import. It is certainly the case that the formal ideas of model, consistency, and independence had not yet been developed at that time. In this text we have taken a little care to set the stage for non-Euclidean geometry.

There are many versions of non-Euclidean geometry. The theories of Beltrami/Klein and Poincaré figure prominently among these. It is a theorem that all two-dimensional non-Euclidean geometries are congruent (see [GRE]). However, some features of the geometry are clearer in one model and other features are clearer in another model. Our decision to present Poincaré's model (below) is one based only on taste. [We note in passing that all models for *Euclidean geometry* are congruent as well. This is proved by way of the Cartesian coordinate system. See [GRE] for details.]

At the risk of overstating the case, let us repeat what we are about to do. We shall not present any new axioms or new ideas. We shall instead present a new notion of "line" that satisfies **P1**, **P2**, **P4** (Axiom **P3** is irrelevant for the present discussion) but does not satisfy **P5**. The notion of "point" will remain the same. These new lines cannot be perfectly arbitrary objects. They must have a certain Euclidean-like structure; that is, they must satisfy the first three axioms. But they will not satisfy the fifth.

The resulting geometry will have a number of peculiar features—peculiar, that is, when viewed from our familiar Euclidean perspective. One of the

---

[2]Here we use the common mathematical notation $\sim$ **P5** to mean "not **P**," or the negation of **P**.

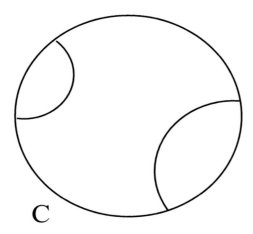

Figure 12.10: A Poincaré line.

reasons that Gauss invoked for not publishing his ideas about non-Euclidean geometry is that he expected that his fellow mathematicians and philosophers would find this all to be an artificial construct of the mind: just a game with ideas that has nothing to do with the real world.

Many times it requires a great deal of time for mathematical ideas to find applications in the real world. It turns out that the model for non-Euclidean geometry that we shall describe, often called *hyperbolic geometry*, is the key to determining the structure of four-dimensional manifolds or surfaces. Four-dimensional manifolds are in turn the language for studying general relativity. Thus, after nearly two centuries, we find that non-Euclidean geometry *is* important in the world around us. However, hyperbolic geometry is more difficult for us to perceive intuitively than standard Euclidean geometry.

The two-dimensional geometry that we now describe takes place not in the infinitely extendible plane but rather inside the unit circle C. Any standard Euclidean point *in the interior* of C is a point for our geometry. We declare a (Poincaré) *line* to be any circular arc that begins and ends on C and is perpendicular to C at both endpoints *or* any standard straight line segment passing through the origin and touching C at both ends. Figures 12.10 and 12.11 illustrate our notion of "line" in this new geometry.

Notice that the second type of line in our geometry—the line segment through the origin—is also perpendicular to C at its endpoints. This is the key geometric fact that characterizes Poincaré lines. Also note that, when we

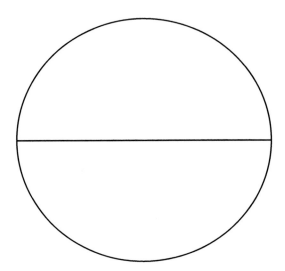

Figure 12.11: Another Poincaré line.

speak of two circular arcs being perpendicular at a point P, then we mean, of course, that their tangent lines are perpendicular at P. An equivalent, but more elementary, way to think about this idea is that the *radii* of the two circular arcs are perpendicular at P: see Figure 12.12.

In what follows we shall engage in an extensive discussion of the artifacts of non-Euclidean geometry. Along the way, we shall describe a variety of geometric constructions. Since this is only an introduction, we shall not provide complete proofs of why the constructions work. We refer the reader to [GRE] for details.

It is not immediately clear that the first type of line even exists. Here is how we can construct them: let x and y be arbitrary points of C that are not on the same diameter. Draw the tangent lines to C at each of these points. These tangents intersect at a unique point z. The circle with center z and passing through x and y will be perpendicular to C at both points of intersection (Figure 12.13). Thus the arc of this circle that lies inside C will be a line in our new geometry.

Why is it true that, through any two points inside C, there passes a line? More fundamentally, why is it true that, through any single point, there passes a line? In order to answer these questions most expeditiously, we need to introduce the concept of *reflection* in the circle C. Let P be a point inside C. Let O denote the center of C. The reflection of P is the unique

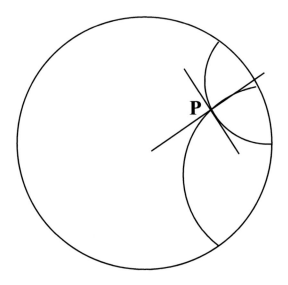

Figure 12.12: Perpendicular radii.

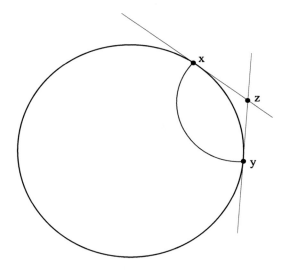

Figure 12.13: A line in the new geometry

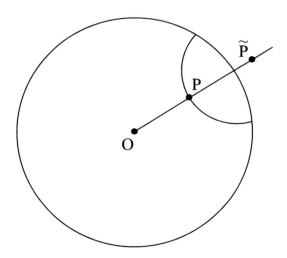

Figure 12.14: Reflection of the point P.

point $\widetilde{P}$ on the ray $\overrightarrow{OP}$ such that $|OP| \cdot |O\widetilde{P}| = 1$. Refer to Figure 12.14.

Now here is the interesting fact: The circle with center $\widetilde{P}$ and passing through P will be orthogonal to C at the points of intersection. Thus the arc of that circle that lies inside C will be a line in our geometry, and it will pass through P. Look again at Figure 12.13. This answers the request for a line through any single given point.

To answer the second question (and establish Axiom **P1**), it is convenient to construct an isomorphism with another model for non-Euclidean geometry—the Beltrami/Klein model. To do this, we use the so-called stereographic projection. As Figure 12.15 suggests, we may project a point P of the Poincaré disc up to a point $\pi(P)$ in the unit sphere by moving along a segment in space that connects the planar point to the north pole N of the sphere. We then project the spherical image vertically down to a point P' in the plane again.

The result of the composition of these two projections is a new planar geometry in which the role of "line" is played by classical straight segments that lie inside a (slightly smaller) circle. Now if we want to know whether two points P and Q in Poincaré's geometry have a (Poincaré) line passing through them, we project the two points up into the sphere and vertically down into the plane again to obtain two points P' and Q' in the Beltrami/Klein geometry. We draw an *ordinary segment* S in the Beltrami/Klein model,

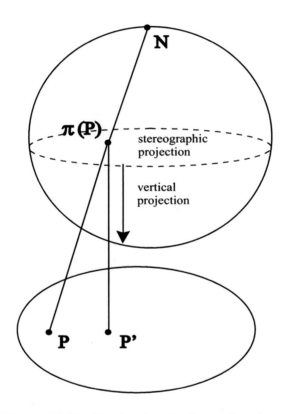

Figure 12.15: Projection to the unit sphere.

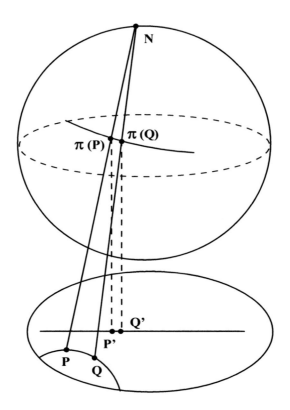

Figure 12.16: A Poincaré line by way of the inverse stereographic projection.

passing through P' and Q', project it up vertically into the sphere, and then apply the inverse stereographic projection to push the circular arc back into the plane. This gives a Poincaré line that passes through the original two points P and Q. Refer to Figure 12.16. Thus we find that Axiom **P1** is satisfied by the (Poincaré) points and (Poincaré) lines in Poincaré geometry.

In Poincaré geometry, the angle between two Poincaré lines at a point of intersection is just declared to be the angle between two classical circular arcs, which is defined to be the classical Euclidean angle between the tangent lines at that point. Thus Axiom **P4** is satisfied.

In order to satisfy Axiom **P2**, we must reconsider our notion of distance. For look at the situation shown in Figure 12.17. If our idea of distance is the ordinary Euclidean idea (measured along a curve with a tape measure for instance), then we would be unable to construct the required point E for these points A, B, C, D. We thus need to modify the notion of distance so

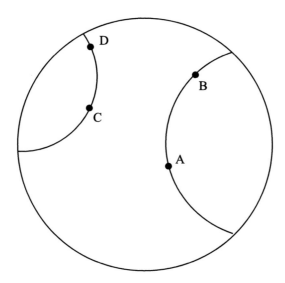

Figure 12.17: A new notion of distance

that points near the boundary are further apart than points near the origin. With this goal in mind, consider a fixed ruler that has length one inch at the origin; the ruler will change scale when it is moved nearer to the boundary of the disc. In other words, our notion of length is dependent on the base point at which we are standing.

The ruler will have length one foot if it is moved sufficiently near the boundary and length one yard if it is moved even nearer. If it is very near indeed to the boundary, then that same yardstick will have length one mile. As the base point tends to the boundary, the length of the ruler will approach infinity.

How can this be? We define the *Poincaré distance* between two points x and y in the unit disc as follows: let P and Q be the endpoints of the Poincaré line through x and y lying on the circle C as in Figure 12.18. We define

$$\rho(x, y) = \left| \log \left(\frac{|xQ||yP|}{|yQ||xP|} \right) \right|.$$

Here $|xP|$ stands for the ordinary Euclidean distance of x to P and similarly for the other expressions.

In order to get a feeling for this notion of distance, let ρ denote the new Poincaré distance. We calculate $\rho(O, X)$, where O is the center of the circle

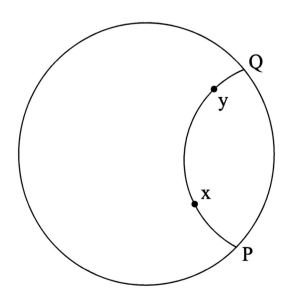

Figure 12.18: Endpoints of a Poincaré line.

and $X = (r, 0)$ with $0 < r < 1$. Of course the Poincaré line through O and X is just the ordinary horizontal diameter segment of C. Now P is the point $(-1, 0)$ and Q is the point $(1, 0)$. Then

$$
\begin{aligned}
|OQ| &= 1 \\
|XP| &= 1 + r \\
|OP| &= 1 \\
|XQ| &= 1 - r.
\end{aligned}
$$

As a result,

$$
\rho(O, X) = \log\left(\frac{1+r}{1-r}\right).
$$

Notice that, as r tends to 1 from the left, the value of $\rho(O, X)$ tends to infinity. In particular, it must be that points out near 1 are getting further and further from the origin and also further and further apart.

An interesting feature of the notion of distance that we have introduced is that it makes the unit disc much like the Euclidean plane equipped with the ordinary Euclidean metric: that is, the unit disc with the Poincaré metric has infinite extent in all directions. In particular, the disc is *complete* in the Poincaré metric in the sense that any sequence of points that gets closer and

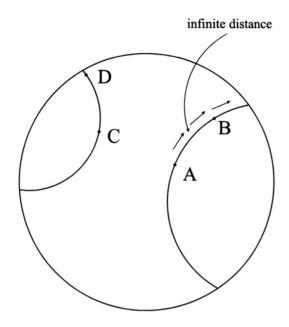

infinite distance

Figure 12.19: Infinite distance to the boundary.

closer together (is *Cauchy*) in the Poincaré metric in the disc actually has a limit in the disc. It is also the case that, just as lines are the curves of shortest length in the Euclidean metric in the plane, so it is that Poincaré lines are the curves of shortest length for the Poincaré metric in the unit disc.

Now return to Figure 12.18. The distance along the curve through A and B and out to the boundary is infinity (Figure 12.19). So there is plenty of room to construct E as in Axiom **P2** (Figure 12.20).

Let us cut this preliminary discussion short and finally consider the crucial Axiom **P5**. First, what does it mean for two lines to be parallel? By definition, they are parallel if they do not intersect. Figure 12.21 exhibits some parallel lines. Now look at Figure 12.22: it exhibits a line ℓ and a point x not on that line. Figure 12.23 shows two distinct Poincaré lines passing through x that are parallel to ℓ. In fact, there are infinitely many such lines! Thus we see that Axiom **P5** fails for Poincaré geometry.

Space limitations prevent us from exploring this new hyperbolic (non-Euclidean) geometry in any depth. It contains many beautiful subtleties. For example, all triangles in non-Euclidean geometry have angles summing to less than 180°. There are no rectangles (quadrilaterals with four right

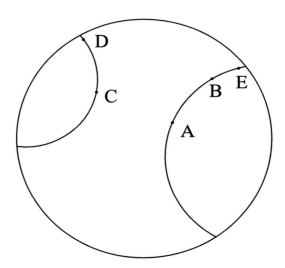

Figure 12.20: Euclid's second axiom in the Poincaré setting.

angles) in non-Euclidean geometry. Perhaps most surprising is that two similar triangles must be congruent: that is, if their angles are the same, then
their side lengths must also be the same. Thus, in particular, photography
would not work very well in non-Euclidean geometry: a lens that shrunk or
expanded objects would perforce have to change angles and therefore would
also distort them.

Exercise: There are no squares in non-Euclidean geometry. And if two
triangles are similar, then they are congruent. Explain why these phenomena
are true.

12.4   Archimedes and the Calculation of the Area of the Circle

12.4.1   The Genius of Archimedes

Archimedes (287 B.C.E.–212 B.C.E.) was born in Syracuse, Sicily. His father
was Phidias, the astronomer. Archimedes developed into one of the most
gifted, powerful, and creative mathematicians who ever lived. You can read
about the life of Archimedes in [AAB] and [DIJ].

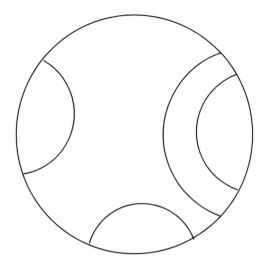

Figure 12.21: Parallel lines in the Poincaré geometry.

One of Archimedes's achievements was to develop methods for calculating areas and volumes of various geometric figures. We shall imitate one of Archimedes's techniques—the method of exhaustion that he learned from Eudoxus (408 B.C.E–355 B.C.E.)—to approximate the area inside a circle to any desired degree of accuracy. This gives us a method for in turn approximating the value of π. It can be said that Archimedes turned the method of exhaustion to a fine art, and that some of his calculations were tantamount to the foundations of integral calculus (which was actually not fully developed until nearly 2000 years later).

Archimedes grew up in privileged circumstances. He was closely associated with, and perhaps even related to, Hieron, King of Syracuse; he was also friends with Gelon, son of Hieron. He studied in Alexandria and developed there a relationship with Conon of Samos; Conon was someone whom Archimedes admired as a mathematician and cherished as a friend.

When Archimedes returned from his studies to his native city he devoted himself to pure mathematical research. During his lifetime, he was regularly called upon to develop instruments of war in the service of his country. And he was no doubt better known to the populace at large, and also appreciated more by the powers that be, for that work than for his pure mathematics. Among his other creations, Archimedes is said to have created (using his understanding of leverage) a device that would lift enemy ships out of the

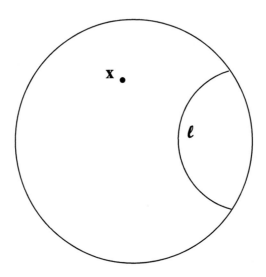

Figure 12.22: Failure of the parallel postulate in Poincaré geometry.

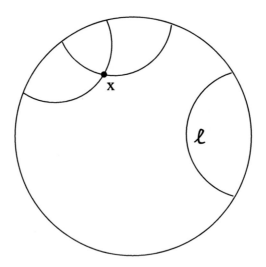

Figure 12.23: Another look at the failure of the parallel postulate.

water and overturn them. Another of his creations was a mirror that would reflect the sun and set enemy ships afire. Archimedes himself set no value on these contrivances, and declined even to leave any written record of them.

Perhaps the most famous story about Archimedes concerns a crown that was specially made for his friend King Hieron. It was alleged to be manufactured of pure gold, yet Hieron suspected that it was actually part silver. Archimedes puzzled over the proper method to determine whether this was true (without modifying or destroying the crown!). Then, one day, as Archimedes was stepping into his bath, he observed the water running over and had an inspiration. He determined that the excess of bulk that would be created by the introduction of silver into the crown could be measured by putting the crown and equal weights of gold and silver separately into a vessel of water—and then noting the difference of overflow. If the crown were pure gold, then it would create the same amount of overflow as the equal weight of gold. If not, then there was silver present.

Archimedes is said to have been so overjoyed with his new insight that he sprang from his bath—stark naked—and ran home down the middle of the street shouting *"Eureka! Eureka!"*, which means "I have found it! I have found it!" To this day, in memory of Archimedes, people cry Eureka! to celebrate a satisfying discovery.

Another oft-told story of Archimedes concerns his having said to Hieron, "Give me a place to stand and I will move the Earth." What Archimedes meant by this bold assertion is illustrated in Figure 12.24. Archimedes was one of the first to study and appreciate the power of levers. He realized that a person of modest strength could move a very great weight with the assistance of the leverage afforded by a very long arm. Not fully understanding this principle, Hieron demanded of Archimedes that he give an illustration of his ideas. And thus Archimedes made his dramatic claim. As a practical illustration of the idea, Archimedes arranged a lever system so that Hieron himself could move a large and fully laden ship.

One of Archimedes's inventions that lives on today is a water screw that he devised in Egypt for the purpose of irrigating crops. The same mechanism is used now in electric water pumps as well as hand-powered pumps in third world countries.

Archimedes died during the capture of Syracuse by the troops of Marcellus in 212 B.C.E. Even though Marcellus gave explicit instructions that

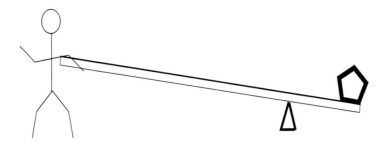

Figure 12.24: The power of a lever.

neither Archimedes nor his house were to be harmed, a soldier became en-
raged when Archimedes would not divert his attention from his mathematics
and obey an order. Archimedes is reported to have said sternly to the soldier,
"Do not disturb my circles!" Thus Archimedes fell to the sword.

Next we turn our attention to Archimedes's study of the area of a circle.

12.4.2   Archimedes's Calculation of the Area of a Circle

Begin by considering a regular hexagon with side length 1 (Figure 12.25).
We divide the hexagon into triangles (Figure 12.26). Notice that each of
the central angles of each of the triangles must have measure $360°/6 = 60°$.
Since the sum of the angles in a triangle is $180°$, and since each of these
triangles certainly has two equal sides and hence two equal angles, we may
now conclude that all the angles in each triangle have measure $60°$. See
Figure 12.27.

But now we may use the Pythagorean theorem to analyze one of the
triangles. We divide the triangle in two—Figure 12.28. Thus the triangle is
the union of two right triangles. We know that the hypotenuse of one of these
right triangles—which is the same as the side of one of the six equilateral
triangles—is 1 and the base is $1/2$. Thus the Pythagorean theorem tells
us that the height of the right triangle is $\sqrt{1^2 - (1/2)^2} = \sqrt{3}/2$. We may
conclude then that the area of this right triangle, as shown in Figures 12.26
and 12.27, is

$$A(T) = \frac{1}{2} \cdot (\text{base}) \cdot (\text{height}) = \frac{1}{2} \cdot \frac{1}{2} \cdot \frac{\sqrt{3}}{2} = \frac{\sqrt{3}}{8}.$$

Therefore the area of the full equilateral triangle, with all sides equal to 1, is

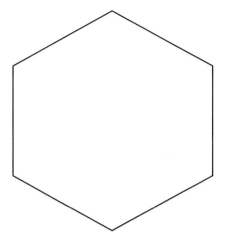

Figure 12.25: A regular hexagon.

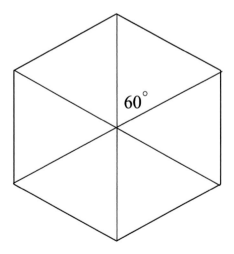

Figure 12.26: The hexagon divided into triangles.

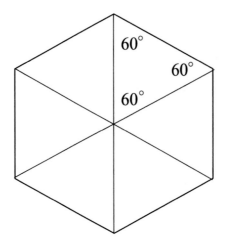

Figure 12.27: The angle measure in each triangle.

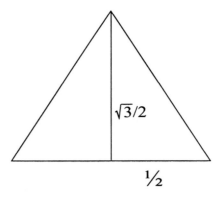

Figure 12.28: Division of the triangle.

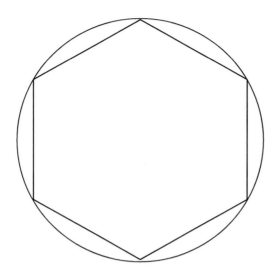

Figure 12.29: Area inside the hexagon approximates the area inside the circle.

twice this or $\sqrt{3}/4$.

Now of course the full regular hexagon is made up of six of these equilateral triangles, so the area inside the hexagon is

$$A(H) = 6 \cdot \frac{\sqrt{3}}{4} = \frac{3\sqrt{3}}{2}.$$

We think of the area inside the regular hexagon as being a crude approximation to the area inside the circle: Figure 12.29. Thus the area inside the circle is very roughly the area inside the hexagon. Of course we know from other considerations that the area inside this circle is $\pi \cdot r^2 = \pi \cdot 1^2 = \pi$. Thus, putting our ideas together, we find that

$$\begin{aligned} \pi &= \text{(area inside unit circle)} \\ &\approx \text{(area inside regular hexagon)} \\ &= \frac{3\sqrt{3}}{2} \approx 2.598\ldots \end{aligned}$$

It is known that the true value of π is $3.14159265\ldots$. So our approximation is quite crude. The way to *improve* the approximation is to increase the number of sides in the approximating polygon. In fact what we shall do is *double* the number of sides to 12. Figure 12.30 shows how we turn one side into two sides; doing this six times creates a regular 12-sided polygon.

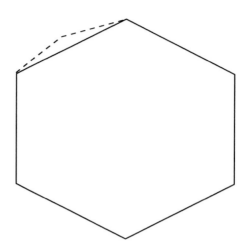

Figure 12.30: Increasing the number of sides in the inscribed polygon.

Notice that we create the regular 12-sided polygon (a dodecagon) by adding small triangles to each of the edges of the hexagon. Our job now is to calculate the area of the twelve-sided polygon. Thus we need to calculate the lengths of the edges. Examine a blown-up picture of the triangle that we have added (Figure 12.31). We use the Pythagorean theorem to calculate the length x of a side of the new dodecagon. It is

$$x = \sqrt{\left(\frac{1}{2}\right)^2 + \left(1 - \frac{\sqrt{3}}{2}\right)^2} = \sqrt{\frac{1}{4} + \left(1 - \sqrt{3} + \frac{3}{4}\right)} = \sqrt{2 - \sqrt{3}}.$$

Now let us focus attention on the dodecagon, divided into twelve isosceles triangles (Figure 12.32). We have just calculated that each side of the dodecagon has length $\sqrt{2 - \sqrt{3}}$. If we can calculate the area of each of the congruent subtriangles, then we can obtain the area of the entire dodecagon (by multiplying by 12). Examine Figure 12.33. This is one of the 12 triangles that makes up the dodecagon. It has base $\sqrt{2 - \sqrt{3}}$. Each of the two sides has length 1. Thus we may use the Pythagorean theorem to determine that the *height* of the triangle is

$$h = \sqrt{1^2 - \left(\frac{\sqrt{2 - \sqrt{3}}}{2}\right)^2} = \sqrt{1 - \frac{2 - \sqrt{3}}{4}} = \sqrt{\frac{2 + \sqrt{3}}{4}}.$$

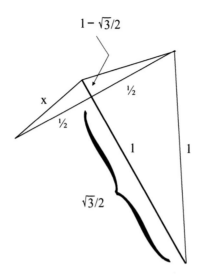

Figure 12.31: The triangle that we have added.

We conclude that the area of the triangle is

$$A(T) = \frac{1}{2} \cdot (\text{base}) \cdot (\text{height}) = \frac{1}{2} \cdot \sqrt{2 - \sqrt{3}} \cdot \sqrt{\frac{2 + \sqrt{3}}{4}} = \frac{\sqrt{4 - 3}}{4} = \frac{1}{4}.$$

Hence the area of the dodecagon is

$$A(D) = 12 \cdot \frac{1}{4} = 3.$$

Examining Figure 12.33, and thinking of the area inside the dodecagon as an approximation to the area inside the unit circle, we find that

$$\pi = (\text{area inside unit circle}) \approx (\text{area inside regular dodecagon}) = 3.$$

This is obviously a better approximation to π than our first attempt. At least we now have the "3" right! Now let us do one more calculation in an attempt to improve the estimate. After that we will seek to find a pattern in these calculations.

Now we consider a regular 24-sided polygon (an icositetragon)—see Figure 12.34. As before, we construct this new polygon by erecting a small triangle over each side of the dodecagon. See Figure 12.35. We examine

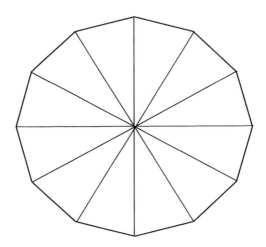

Figure 12.32: The dodecagon divided into triangles.

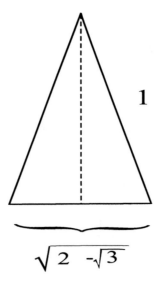

$$\sqrt{2 - \sqrt{3}}$$

Figure 12.33: The area of the dodecagon.

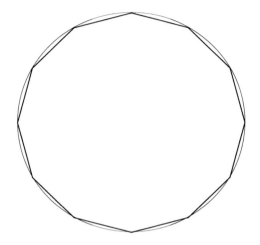

Figure 12.34: Approximation to the area inside a circle.

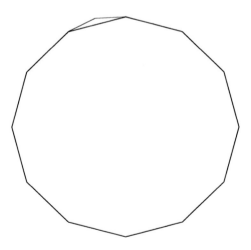

Figure 12.35: The icositetragon.

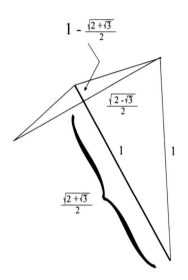

Figure 12.36: Adding a triangle.

a blowup (Figure 12.36) of one of these triangles, just as we did above for the dodecagon. We first solve the right triangle with base $\sqrt{2 - \sqrt{3}}/2$ and hypotenuse 1—using the Pythagorean theorem, of course—to find that it has height $\sqrt{2 + \sqrt{3}}/2$. Then we see that the smaller right triangle has base $1 - \sqrt{2 + \sqrt{3}}/2$ and height $\sqrt{2 - \sqrt{3}}/2$. Thus, again by the Pythagorean theorem, the hypotenuse of the small right triangle is $\sqrt{2 - \sqrt{2 + \sqrt{3}}}$.

But the upshot is that the icositetragon is made up of isosceles triangles, as in Figure 12.36, having base $\sqrt{2 - \sqrt{2 + \sqrt{3}}}$ and side length 1. We may divide the triangle into two right triangles, as indicated in the figure, and then solve one of the right triangles using the Pythagorean theorem. The solution is that the height of this right triangle is $\sqrt{2 + \sqrt{2 + \sqrt{3}}}/2$. Altogether, then, the area of the triangle, which is one twenty-fourth of the polygon, is

$$
\begin{aligned}
A(T) \;&=\; \frac{1}{2} \cdot (\text{base}) \cdot (\text{height}) \\[2mm]
&=\; \frac{1}{2} \cdot \sqrt{2 - \sqrt{2 + \sqrt{3}}} \cdot \frac{\sqrt{2 + \sqrt{2 + \sqrt{3}}}}{2} \\[2mm]
&=\; \frac{\sqrt{2 - \sqrt{3}}}{4}.
\end{aligned}
$$

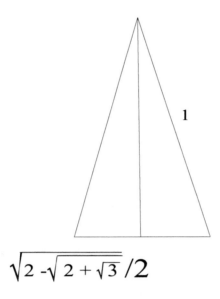

$$\sqrt{2 - \sqrt{2 + \sqrt{3}}}\,/2$$

Figure 12.37: The icositetragon broken up into triangles.

We conclude that the area of the 24-sided regular polygon is

$$A(P) = 24 \cdot \frac{\sqrt{2 - \sqrt{3}}}{4} = 6\sqrt{2 - \sqrt{3}}\,.$$

Examining Figure 12.37, and thinking of the area inside the dodecagon as an approximation to the area inside the unit circle, we find that

$$\pi = (\text{area inside unit circle}) \approx (\text{area inside regular 24-gon}) \approx 3.1058\,.$$

We see that, finally, we have an approximation to π that is accurate to one decimal place. See Figure 12.38.

Of course the next step is to pass to a polygon of 48 sides. We shall not repeat all the steps of the calculation but just note the high points. First, we construct the regular 48-gon by placing small triangles along each of the edges of the dodecagon. See Figure 12.39. Now, once again, we must (blowing up the triangle construction) examine a figure like 12.40. The usual calculation shows that the side of the small added triangle has length $\sqrt{2 - \sqrt{2 + \sqrt{2 + \sqrt{3}}}}$. Thus we end up examining a new isosceles triangle, which is 1/48th of the 48-sided polygon. See Figure 12.40.

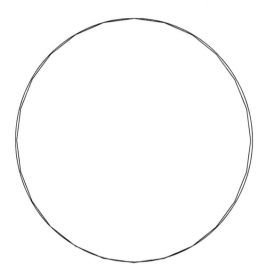

Figure 12.38: Approximation of the area inside a circle.

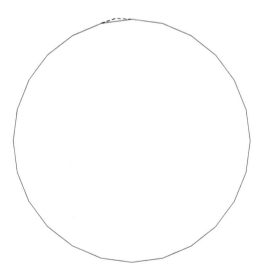

Figure 12.39: The regular 48-gon.

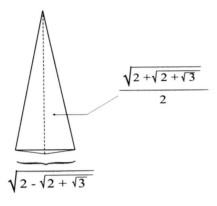

Figure 12.40: Adding a triangle.

The usual calculations, just as we did for the polygons having 6 or 12 or 24 sides, show that this new triangle has base $\sqrt{2 - \sqrt{2 + \sqrt{2 + \sqrt{3}}}}$ and height $\sqrt{2 + \sqrt{2 + \sqrt{2 + \sqrt{3}}}}/2$. Thus the area is

$$
\begin{aligned}
A(T) &= \frac{1}{2} \cdot (\text{base}) \cdot (\text{height}) \\
&= \frac{1}{2} \cdot \sqrt{2 - \sqrt{2 + \sqrt{2 + \sqrt{3}}}} \\
&\qquad \cdot \sqrt{2 + \sqrt{2 + \sqrt{2 + \sqrt{3}}}}/2 \\
&= \frac{\sqrt{2 - \sqrt{2 + \sqrt{3}}}}{4} .
\end{aligned}
$$

The polygon comprises 48 such triangles, so the total area of the polygon is

$$
A(P) = 48 \cdot \frac{\sqrt{2 - \sqrt{2 + \sqrt{3}}}}{4} = 12\sqrt{2 - \sqrt{2 + \sqrt{3}}}.
$$

See Figure 12.41. Thinking of the area inside the 48-sided regular polygon

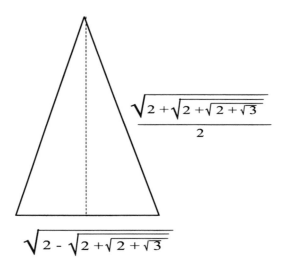

Figure 12.41: The area of a triangle inside the 48-gon.

as an approximation to the area inside the unit circle, we find that

$$
\begin{aligned}
\pi &= \text{(area inside unit circle)} \\
&\approx \text{(area inside 48-sided regular polygon)} \\
&\approx 3.1326\,.
\end{aligned}
$$

This is obviously a better approximation to π than our last three attempts. It is accurate to one decimal place, and the second decimal place is close to being right.

And now it is clear what the pattern is. The next step is to examine a regular polygon with 96 sides. The usual calculations will show that this polygon breaks up naturally into 96 isosceles triangles, and each of these triangles has area

$$
A(T) = \frac{\sqrt{2 - \sqrt{2 + \sqrt{2 + \sqrt{3}}}}}{4}\,.
$$

Thus the area of the polygon is

$$
A(P) = 96 \cdot \frac{\sqrt{2 - \sqrt{2 + \sqrt{2 + \sqrt{3}}}}}{4} = 24 \cdot \sqrt{2 - \sqrt{2 + \sqrt{2 + \sqrt{3}}}}\,.
$$

We then see that

$$\pi = \text{(area inside unit circle)}$$
$$\approx \text{(area inside 96-sided regular polygon)}$$
$$\approx 3.13935.$$

This is certainly an improved approximation to the true value of π, which is 3.14159265.....

The next regular polygon in our study has 192 sides. It breaks up naturally into 192 isosceles triangles, each of which has area

$$A(T) = \frac{\sqrt{2 - \sqrt{2 + \sqrt{2 + \sqrt{2 + 3}}}}}{4}.$$

Thus the area of the regular 192-gon is

$$A(P) = 192 \cdot \frac{\sqrt{2 - \sqrt{2 + \sqrt{2 + \sqrt{2 + 3}}}}}{4} = 48 \cdot \sqrt{2 - \sqrt{2 + \sqrt{2 + \sqrt{2 + 3}}}}.$$

We then see that

$$\pi = \text{(area inside unit circle)}$$
$$\approx \text{(area inside 192-sided regular polygon)}$$
$$\approx 3.14103.$$

This new approximation of π is accurate to nearly three decimal places.

Archimedes himself considered regular polygons with nearly 500 sides. His method did not yield an approximation as accurate as ours. But, historically, it was one of the first estimations of the size of π.

We would be remiss not to point out the pattern in the forms that $A(T)$ takes as the number of sides of the embedded polygon doubles. An examination of this pattern should enable you to predict, without actually doing any calculations, what $A(T)$ will be for an embedded regular polygon with 384 sides.

For You to Try: The method that we have used here to estimate the area inside a circle can also be used to estimate the area under a parabola (such as $y = -x^2 + 1$) and above the x-axis. It can also be used for 3-dimensional volumes, such as estimating the area inside a sphere.

Exercise: Use the method of Archimedes to calculate the area under the graph of $f(x) = x^3$, above the x-axis, and between the vertical lines $x = 1$ and $x = 2$.

Chapter 13

Concluding Remarks

13.1 The Final Word

In its very early days, mathematics was motivated by land management issues. There was virtually no theory involved. Everything was pragmatic and epistemological.

But the great thing about mathematical ideas is that they live forever. Today, 5,000 years later, mathematics is still used to analyze land management issues. And the ideas that were developed millennia ago are still valid today.

The good news is that mathematics has developed over time in startling and diverse ways. Today mathematics is used for DNA sequencing, for cryptography, for Internet routing, for auto body design, for image compression, for geometric visualization, and for many hundreds of other exciting applications. The number of new mathematical techniques that are developed every week is bewildering and impressive.

The purpose of this book has been to help you to see and understand what mathematics is and how it can be used. The examples that we presented are accessible, meaningful, exciting, and useful. They should give you food for further thought, and perhaps even motivate you to do further reading in the subject.

You may not decide to become a mathematician, but we hope that you are now more open to mathematical ideas, and more aware of the role of mathematics in our lives. We hope that your critical thinking skills have been enhanced, and your ability to understand and analyze problems has been developed.

And we wish you a pleasant journey.

References

[**AAB**] A. Aaboe, *Episodes from the Early History of Mathematics*, Mathematical Association of America, Washington, D.C., 1964.

[**ASW**] T. Aste and D. Weaire, *The Pursuit of Perfect Packing*, Institute of Physics Publishing, Bristol, U.K., 2000.

[**BAS**] Thomas A. Bass, *The Eudaemonic Pie*, Universe Publishing, Houghton-Mifflin, New York, 1985.

[**BCG**] Elwyn Berlekamp, John Horton Conway, and Richard Guy, *Winning Ways for Your Mathematical Plays*, Academic Press, New York, 1982.

[**BES1**] A. S. Besicovitch, A general form of the covering principle and relative differentiation of additive functions, I, *Proc. Cambridge Philos. Soc.* 41(1945), 103–110.

[**BES2**] A. S. Besicovitch, A general form of the covering principle and relative differentiation of additive functions, II, *Proc. Cambridge Philos. Soc.* 42(1946), 205–235.

[**CON**] John H. Conway, *On Numbers and Games*, Academic Press, San Francisco, CA, 1976.

[**CGS**] John H. Conway, C. Goodman-Strauss, and Neil J. A. Sloane, Recent progress in sphere packing, *Current Developments in Mathematics*, 1999 (Cambridge, MA), 37–76, International Press, Somerville, MA, 1999.

[**CUN**] F. Cunningham, The Kakeya problem for simply connected and for star-shaped sets, *Am. Math. Monthly* 78(1971), 114–129.

[**CRCCP**] William M. Haynes, ed., *CRC Handbook of Chemistry and Physics*, 96th ed., CRC Press, Boca Raton, FL, 2015.

[**CRC**] Daniel Zwillinger, ed., *CRC Standard Mathematical Tables and Formulas*, 30th ed., CRC Press, Boca Raton, FL, 1996.

[**DIJ**] E. J. Dijksterhuis, *Archimedes*, Princeton University Press, Princeton, NJ, 1987.

[**DRO1**] Michael Drosnin, *The Bible Code*, Touchstone Books, New York, 1998.

[**DRO2**] Michael Drosnin, *The Bible Code, II*, Penguin Books, New York, 2003.

[**ERD**] P. Erdős, On the fundamental problem of mathematics, *Amer. Math. Monthly* 79(1972), 149–150.

[**FTO**] L. Fejes Tóth, On close-packings of sphere in spaces of constant curvature, *Publ. Math. Debrecen* 3(1953), 158–167.

[**FHS**] A. Finkelstein, N. Hendren, and M. Shepard, Subsidizing health insurance for low-income adults: evidence from Massachusetts, `https://scholar.harvard.edu/files/hendren/files/commcare.pdf`

[**GAM**] G. Gamow, *One Two Three ... Infinity*, Dover Books, New York, 1988.

[**GOF**] C. Goffman, And what is your Erdős number?, *Amer. Math. Monthly* 76(1969), 791.

[**GOO**] I. J. Good, The Paradox of Confirmation, 11(1960), 145–149. JSTOR 685588

[**GRE**] M. J. Greenberg, *Euclidean and Non-Euclidean Geometries; Development and History*, 4th ed., W. H. Freeman and Company, New York, 2007.

[**GUZ**] M. de Guzman, *Differentiation of Integrals in* $\mathbb{R}^n$, Springer Lecture Notes, Springer-Verlag, Berlin and New York, 1975.

[**HAL1**] Thomas Hales, The status of the Kepler conjecture, *Math. Intelligencer* 16(1994), 47–58.

[**HAL2**] Thomas Hales, A proof of the Kepler conjecture, *Annals of Math.* 162(2005), 1065–1185.

[**HAL3**] Thomas Hales, Formal proof, *Notices of the American Mathematical Society* 55(2008), 1370–1380.

[**HAN**] B. Han, Global early warning system for infectious diseases, `https://www.sciencedaily.com/releases/2016` `/05/160520101029.htm`

[**HAD**] B. A. Han and J. M. Drake, Future directions in analytics for infectious disease intelligence, `https://www.ncbi.nlm.nih.gov/pmc/articles/PMC5278609/`

[**HPJA**] B. A. Han, A. W. Park, A. E. Jolles, and S. Altizer, Infectious disease transmission and behavioural allometry in wild animals, `onlinelibrary.wiley.com/doi/10.1111/1365-2656.12336` `/abstract`

[**HAR**] G. H. Hardy, *A Mathematician's Apology*, Cambridge University Press, Cambridge, 1967.

[**HOD**] A. Hodges, *Alan Turing : The Enigma*, Simon and Schuster, New York, 1983.

[**HOK**] Richard Hoshino and Ken-ichi Kawarabayashi, Graph theory and sports scheduling, *Notices of the American Mathematical Society* 60(2013), 726–731.

[**HSI1**] Wu-Yi Hsiang, On the sphere packing problem and the proof of Kepler's conjecture, *Internat. J. Math.* 4(1993), 739–831.

[**HSI2**] Wu-Yi Hsiang, Sphere packings and spherical geometry—Kepler's conjecture and beyond, Center for Pure and Applied Mathematics, U. C. Berkeley, July, 1991.

[**HSI3**] Wu-Yi Hsiang, *Least Action Principle of Crystal Formation of Dense Packing Type and Kepler's Conjecture*, World Scientific, River Edge, NJ, 2001.

[**HSI4**] Wu-Yi Hsiang, A rejoinder to T. C. Hales's article: "The status of the Kepler conjecture" *Math. Intelligencer* 16(1994), *Math. Intelligencer* 17(1995), 35–42.

[**KNU1**] Donald Knuth, *The Art of Computer Programming*, vols. 1–4, Addison-Wesley, Upper Saddle River, NJ, 1997–2011.

[**KNU2**] Donald Knuth, The complexity of songs, *SIGACT News* 9(1977), 17–24.

[**KNU3**] Donald Knuth, *Surreal Numbers: How Two Ex-Students Turned on to Pure Mathematics and Found Total Happiness*, Addison-Wesley, Reading, MA, 1974.

[**KOB**] N. Koblitz, *A Course in Number Theory and Cryptography*, Springer-Verlag, New York, 1987.

[**KOL**] Andrey Kolmogorov, *Foundations of Probability Theory*, Chelsea, New York, 1950.

[**KRA1**] Steven G. Krantz, *The Elements of Advanced Mathematics*, 3rd ed., Taylor & Francis, Boca Raton, FL, 2012.

[**LIT**] J. E. Littlewood, *A Mathematician's Miscellany*, Methuen, London, 1953.

[**LOY**] Sam Lloyd and Martin Gardner, *Mathematical Puzzles of Sam Lloyd*, Dover Publications, New York, 1959.

[**MAH**] P. Maher, Inductive logic and the ravens paradox, *Philosophy of Science*, 66(1999), 50–70.

[**MOV**] Oscar Morganstern and John von Neumann, *Theory of Games and Economic Behavior*, Princeton University Press, Princeton, NJ, 1944.

[**PAUL1**] J. A. Paulos, *Innumeracy*, Hill and Wang, New York, 1988.

[**REN**] P. Renz, Thoughts on *Innumeracy*: Mathematics versus the world, *Am. Math. Monthly* 1993, 732-742.

[**ROB**] Abraham Robinson, *Non-Standard Analysis*, North-Holland Publishing, Amsterdam, 1966.

[**SEY**] Paul Seymour, Progress on the four-color theorem, *Proceedings of the International Congress of Mathematicians*, Vol. 1, 2 (Zürich, 1994), 183–195, Birkhäuser, Basel, 1995.

[**THO**] D'Arcy Thompson, *On Growth and Form*, Cambridge University Press, Cambridge, 1952.

[**THOR1**] Edward Thorp, *Beat the Dealer*, Vintage Press, New York, 1962.

[**THOR2**] Edward Thorp and Sheen T. Kassouf, *Beat the Market*, Random House, New York, 1967.

[**TUR**] S. Turing, *Alan M. Turing*, W. Heffer, Cambridge, 1959.

[**WIL**] Rebecca S. Wills, Google's PageRank: The math behind the search engine, `http://www.cems.uvm.edu/~tlakoba/AppliedUGMath /other_Google/Wills.pdf`

Index